Astronomers' Universe

W0230480

Series editor

Martin Beech, Campion College, The University of Regina, Regina,
Saskatchewan, Canada

More information about this series at http://www.springer.com/series/6960

Jonathan Powell

Cosmic Debris

What It Is and What We Can Do About It

 Springer

Jonathan Powell
Ebbw Vale, Gwent
UK

ISSN 1614-659X   ISSN 2197-6651 (electronic)
Astronomers' Universe
ISBN 978-3-319-51015-6   ISBN 978-3-319-51016-3 (eBook)
DOI 10.1007/978-3-319-51016-3

Library of Congress Control Number: 2017934208

Printed on acid-free paper

This Springer imprint is published by Springer Nature
The registered company is Springer International Publishing AG
The registered company address is: Gewerbestrasse 11, 6330 Cham, Switzerland

Contents

About the Author

Jonathan Powell is a native of South Wales. With an interest in astronomy and related subjects since the late 1970s, Jonathan has written for two of the United Kingdom's largest astronomical magazines, Astronomy Now and The Sky at Night. Jonathan has been a freelance radio broadcaster on astronomy since 1985, chiefly for regional BBC output. During this time he has had two monthly slots, firstly on BBC Radio Gwent, then a three year run on BBC Radio Wales. Jonathan also took part in a BBC Radio 4 documentary series on meteorites. He has held the position of secretary for one of the local astronomical societies, with involvement for a number of years as local officer for the British Astronomical Association's Campaign for Dark Skies.

A former member of the Association for Astronomy Education—which encouraged and promoted astronomy among the young—Powell also headed up 'SpaceQuest,' an educational lecture tool, which toured schools with a two-hour presentation on astronomy. Jonathan has also written two monthly astronomy columns, firstly for a local newspaper which ran for seven years back in the 1980's entitled *Stargazing*, and more recently for a regional newspaper, the South Wales Argus, with a column entitled *The Night Sky*. Jonathan has also written a book on Welsh castles entitled *Fortress Wales*.

1. Defining Cosmic Debris

First Light

The cycle begins and the clock is ticking. The potent force that is to encapsulate all that we know is reaching out, expanding, pioneering its first great exhalation of breath into the great void—a breath that will encompass all, as the bud turns to flower and opens its petals, spreading light where once only darkness existed. The seconds pass, the minutes accumulate, days turn into weeks, months into years, and the giant exhalation continues—the dawn of time and the first flicker of consciousness.

Difficult to comprehend but undeniable by its very presence, it all had to start somewhere at some point and, with everything that starts, there is an end. The mayfly has but 24 h in its life cycle from birth to death, with some of the 2500 known species of mayfly living for an even shorter time, barely a few hours. But these hours are all relative, as the length of time spent is as full in the life of a mayfly as the 200-year potential lifespan of a bowhead whale. However, time for both and for all, as we know, will end.

Physics dictates that time will at some point literally run out, simply cease to be. The ticking clock will ultimately stop. Could it be at this point, when time does finally end, that another Big Bang sets it all in motion again? And, if this is the case, like life itself, is the world in which we live merely made up of endless cycles?

Is it one great exhalation followed by another great inhalation, drawing all that we know back to one starting point, so another such exhalation can start again?

From the dawn of time these questions have been posed, and will continue to be posed, but even with our greater understanding of the mechanics of the universe, there will always be questions that are simply unanswerable. By the same token, to know everything can be considered rather dangerous, so it is probably quite reassuring that we don't know everything. It is perhaps an uncomfortable imbalance for some, but a great leveler for others.

© Springer International Publishing AG 2017
J. Powell, *Cosmic Debris*, Astronomers' Universe,
DOI 10.1007/978-3-319-51016-3_1

This 'not knowing' also acts as a catalyst to push humans to want to know more—the very driver behind climbing the highest mountain, diving to the deepest depths of the oceans, indeed, reaching for the stars. The thirst for knowledge will never end, making the journey towards unraveling life's puzzles more inviting, with perhaps the discovery not always living up to the expectation, the journey itself being more of a revelation than the ultimate findings.

In order to understand the place of cosmic debris in our lives, we first have to understand the known parameters in which it exists (Fig. 1.1).

Fig. 1.1 Our Universe (courtesy of NASA)

The universe is the totality of existence, all-encompassing, from the vastness of a galaxy to the smallest subatomic particles. All matter and all energy is accounted for within its vast scope, including dark matter and dark energy.

The visible part of the universe, which includes our own Earth, our own Sun, other stars, and the most distant of galaxies, represents this visible universe, made from protons, neutrons and electrons, clustered together into atoms. The rest of the universe is accounted for by an invisible substance known as dark matter, and by a force that repels gravity, known as dark energy.

The Matter of Darkness

Percentage-wise, dark energy has the majority share in the universe, comprising some 70%, with dark matter accounting for 25%.

Fritz Zwicky (1898–1974), a Swiss astronomer at the California Institute of Technology (Caltech), stumbled across the gravitational effects of dark matter in the early 1930s. Zwicky, a bold and visionary scientist, was far ahead of his time in conceiving of the existence of dark matter.

At the time, Zwicky was studying how galaxies move within the Coma cluster, a large collection of galaxies located in the constellation of Coma Berenices. Over 1000 identified galaxies dwell within the cluster, spanning about 2° in the night sky. All told, the cluster could contain as many as 10,000 galaxies (each housing billions of stars), representing one of the richest known clusters (Fig. 1.2).

Zwicky was interested in how gravity affected the movements of galaxies within the cluster. The galaxies within the cluster have no central heavy object, like our Sun, to gravitate around, and thus follow more complicated orbits. How were the complicated orbits controlled, though?

Zwicky's observations of supernovae in distant galaxies, made from Caltech's Mount Wilson Observatory, laid the foundation of his theoretical work.

Having persuaded Caltech to build an 18-inch Schmidt telescope that could capture large numbers of galaxies in a single

Fig. 1.2 Coma Berenices, Coma Cluster (courtesy of NASA)

wide-angle photograph, Zwicky was able to carry out work in greater detail. As he detected the presence of supernovae (another of his discoveries) in ever-more distant galaxies, he realized that most galaxies group in clusters, like the Coma cluster. Careful measurements of the light from these galaxy clusters led Zwicky to suggest the existence of dark matter.

This dark matter remains completely invisible to human detection, and has never been directly observed. Its non-interaction with baryonic matter, coupled with the dark matter being invisible to light, electromagnetic radiation, and any other form of recognition, makes the perception of its very existence somewhat of a conundrum. How do we know it is there when we can't see it or even detect it?

Despite a lack of tangible evidence, there is confidence in the existence of dark matter due to the gravitational effects placed on galaxy clusters. These clusters, potentially containing thousands of objects, are all bound together by gravity.

Apart from the gravitational force that binds the galaxies together as if in a cluster, what evidence is there to suggest an external force, or another pull, being exerted on them from elsewhere? Physics dictates that stars at the edge of a spinning, spiral galaxy should travel a great deal slower than the stars situated near to the galactic center. From the edge of the galaxy, slower speeds steadily increase in pace as you approach the center, where the galaxy's visible matter is concentrated. Observations, though, have contradicted this, with findings revealing that stars orbit the galactic center at the same rate right across the galaxy's disc, a steady rotation, with only minor discrepancies. It is a reasonably safe conclusion that something is affecting the stars' motions, perhaps a shroud engulfing the galaxy, an unseen mass exerting influence—indeed, dark matter.

Although scientists can't see it, the presence of dark matter may account for certain optical illusions that astronomers have witnessed. Pictures of distant galaxies have been documented as displaying curious rings, strange arcs of light. These bizarre apparitions could be explained away by the presence of dark matter. If light from these distant galaxies is being distorted, it could be possible that large dark matter clouds are present in between the observer and the galaxy, with the matter producing a magnifying effect on the galaxy—gravitational lensing.

Many theories exist as to what dark matter actually is. Firstly, apart from the arguments around its percentage within the universe—some estimates putting it at 80% rather than 70%—it could well be a collection of rather exotic particles that refuse to interact with normal matter or light, but which can nevertheless still exert a gravitational pull on objects.

If dark matter is composed of non-baryonic matter, which a portion of scientists subscribe to as being a likely theory, then the matter could be composed of WIMPS: weakly interacting massive particles. These particles have 10 to 100 times the mass of a proton, but because they are 'weak' their interactions with matter that is considered 'normal' makes them difficult to detect, indeed, invisible to detection. These WIMPS also harbor TeV-scale masses, natural by-products of theories of supersymmetry (SUSY), or possible extra dimensions. The theory of supersymmetry predicts new families of particles interacting very weakly with ordinary matter. The lightest of these supersymmetry particles could be a dark matter particle.

Alongside WIMPS, particle physics also proposes, from the same stable, another possible candidate, the presence of axions—

hypothetical new particles associated with quantum chromodynamics (QCD).

Another theory carrying equal weight for consideration is the presence of neutralinos—massive hypothetical particles. These particles are heavier and slower than neutrinos, and, like dark matter, have never been seen.

By far the most radical theory is that the laws of gravity, while comfortably explaining everything up until dark matter, do not account for this phenomenon, and that a full revision of gravitational laws needs to be undertaken.

From Matter to Energy

Dark matter is one puzzle, but what of dark energy? Its discovery in the 1990s came as a shock to scientists who had assumed that the attractive force of gravity would slow down the expansion of the universe in time, with the clock continuing to tick, but at a much slower and reduced rate. However, on attempting to measure the rate deceleration, scientists discovered that the expansion was speeding up. The clock was increasing in pace, time passing with more fluidity. Two independent teams came up with the same result, with one scientist likening the find to throwing a set of keys in the air and expecting them to fall back down, only to see them fly straight up and towards the ceiling.

One explanation for dark energy is that is simply a property of space itself, something that exists, like dark matter, but that has eluded proper classification. In other words, it is all part of the universe and although its function and makeup remain a mystery, the fact that it exists does not. In order to further clarify its existence, 'space' must be seen as being a matrix in which the universe and cosmos are present.

Albert Einstein was the first person to realize that empty space is not nothing. Nothing has to have something within it. It cannot just contain nothing. Einstein discovered that it is possible for more space to come into existence. Einstein's gravity theory contains a cosmological constant, that so-called "empty space" can possess its own energy. Because this energy is a property of space itself, it would not be diluted as space expands. Over time, if more space were to come into existence, more of the energy-of-space would appear. As a result, and explaining the acceleration of the expansion of the universe, this form of energy would cause the universe to expand at an ever-increasing rate.

Another feasible explanation for "empty space" is that, apart from not being empty, it is being occupied with temporary particles. These "virtual" particles, would constantly form, appear, and then disappear. Calculations to fit this theory left more questions than answers, so although it remains a plausible and interesting theory, its potential as an answer falls short of its promise.

Could dark energy be a new kind of energy fluid or field? This fluid would occupy all of space, but as something whose effect on the expansion of the universe is opposite to that of matter and normal energy. Some theorists have named this "quintessence," after the fifth and highest element in ancient and medieval philosophy. This fifth element permeates all nature and is the substance of which celestial bodies are composed. It is more precisely referred to as a scalar field.

We Will Have Order

The cosmos, while complex, is both ordered and seemingly disciplined, with its own natural rhythmic pattern. Rather than a random scattering of objects, there is a structure. For example, one noticeable constant, that of gravity, has neatly ordered the distribution of galaxies throughout the universe, with order placed on the way in which they group together in clusters, a sort of cosmic symmetry. Within a billion light years of Earth, there are swarms of superclusters, groups of groups of galaxies, all tethered by the influence of gravity.

When we look out into the universe, all that we see must be close enough for light to have reached us since the clock starting ticking—since the Big Bang. Because the universe is in a state of expansion, the most distant visible objects are farther away than the material that dates the universe to around 14 billion years old. In reality, due to this expansion, and due to the fact that photons have traveled 45 billion light years to reach us, the visible universe is consequently something in the region of 90 billion light years across. Even that estimation is probably considerably short of the actual size.

However, just to complicate affairs further, the expansion rate of the universe is not constant, with expansion occurring at more or less any speed, including that of being faster than the speed of light.

Therefore, the most distant observable objects we can now see were in fact once much closer to us. During the millions of years that have passed, the universe has shifted distant objects away from us, relocating them to more distant locations. This means that we can now see objects that are even further away than previously believed, expanding our limit of the observable universe.

A Sky Full of Galaxies

With the universe known to be literally full of galaxies, it is unsurprising that the most distant object to be observed is a galaxy. Although this will undoubtedly change over time, it will probably only be another distant galaxy taking its place. Many galaxies have interesting features, with this galaxy, NGC 253 in the southern hemisphere constellation of Sculptor, discovered by Caroline Herschel on September 23, 1783. It is believed that a supermassive black hole exists at the heart of the galaxy with a mass some five million times that of our Sun (Fig. 1.3).

How, then, are we to determine the actual size of the universe? What methods can we use? There are various models to suggest how big the universe should be, given the data to hand, but the very premise on which to start analysis must be around the 'shape' of the universe, whether it be spherical in appearance, flat, or open. If it is spherical, then boundaries can be established and measurements applied; if it is flat or open, then measurements cannot be applied.

One approach used to attempt to measure the universe, assuming that it is of a spherical nature, is to search for the most remote observable object of known size and measure just how big it looks. If it is bigger than it ought to be, the universe can be determined as closed. If the size of the object is correct with no

Fig. 1.3 NGC 253. Enhanced image of the intermediate spiral galaxy (courtesy of NASA)

distortion, the universe must therefore be flat. If the object appears smaller than it ought to be, the universe is open.

Various data analysis, using an averaging technique on the information required, would seem to suggest that the universe is flat. If this is correct, then the universe would be a staggering 250 times bigger than the Hubble volume or Hubble sphere. This is a spherical region of the universe beyond which objects recede from the observer, at a rate greater than the speed of light, due to the expansion of the universe. This concept is both fascinating and troublesome at the same time, as trying to comprehend such vastness is almost impossible.

The most distant light that astronomers can see emanates from cosmic microwave background radiation. This radiation is composed of photons that have traveled towards us from nearly the beginning of the universe, from the Big Bang itself.

Initially, everything was compacted in, compromised and curtailed, meaning that light traveling from any object would not have clear sight to make a direct path to a faraway object, being scattered, or absorbed by another particle. As time passed and the universe expanded, these obstacles to light were slowly removed, allowing a far greater reach than in earlier times.

Therefore, the universe presents us with one massive learning curve, and whereas, piece by piece, the jigsaw is being assembled, there is unlikely to be a day when the final piece is put into place. As our understanding grows of the environment in which we live, so does the uncovering of further questions, which in turn, while often exasperating, affords us a chance to re-examine what we already know, allowing for a pause in the cognitive process.

The world guides us by our senses and, as fickle as we are with some of the understandings of our place in the universe, it is the quest for knowledge that allows us, like the universe, to expand within it, both mentally and physically, as we push back the boundaries of our ever-wakening awareness of our surroundings.

A Bubbly Feel

The universe has a 'feel' to it. Galactic surveys have revealed its texture to be 'bubbly' in nature. A distinct majority of galaxies are found within giant arcs surrounding extensive voids in space, with some measuring more than 300 million light years across. These barren areas of the universe contain virtually nothing; they are literally empty with regards to objects, but literally full with regards to dark matter, which of course, as previously noted, cannot be seen. Overall, it has been determined that the universe is a vast collection of knots, sheets, and filaments, all combining to make up the entire fabric.

The challenge ahead has always been to 'grow' rather than 'expand.' Expansion often potentially infers conquest, taking what we believe to be rightfully ours—indeed, as with the early space race, competing to be first—when really it is the journey to get there, not the arrival.

History dictates that 'firsts' in life denote power, strength, and capability, which may well be true for many of life's rites of passage, but not with space. We know the first into space, the first to orbit the Moon, the first to land on the Moon, and so on, but is a collection of firsts like that really what the rolling comprehension of our universe is all about? There is much more at stake here than merely getting somewhere first, and, in order to fully understand, one has to take time out to realize just how substantial the universe is.

Only a fool fights in a burning house. Whereas two insects may grapple for supremacy over a rotten apple, learning to know one's place in the universe can only encourage a better understanding of what really matters.

Then There's the Debris

In that vastness of space lies the debris. From the smallest grain of sand to a several-kilometer-wide asteroid and beyond, it's all out there.

The various definitions of debris over time have given us a better understanding as to the formation of our Solar System and all that lies within in. Although the notion of debris might encourage the thought of something to be ignored, trivial, not worth mentioning, it is as vital a part of the universe as the largest planet and most distant galaxy.

The truth is that classification and subsequent definitions of debris need to be established and regularly maintained and updated. For the most part, the code of ethics with regard to cosmic debris has been adhered to, but there will always be cases presented for change, one widely publicized recent example being that of Pluto, unceremoniously stripped of its status, indicating that, not just in the world of debris, size does matter.

With the discovery of a dwarf planet in the Kuiper Belt, the decision to reclassify Pluto has perhaps been justified, but it does call into question just how certain decisions are arrived at with regard to classification, and whether or not one day the whole system will have to be overhauled.

The Reclassification of Pluto

At a distance of 9 billion km from the Sun, a ball of rock measuring around 700 km orbits the Sun. Covered with ice and frozen nitrogen, 2015 RR245 makes one circuit of the Sun every 700 years, which gives you an idea of just exactly how far distant it is from Pluto, which orbits the Sun every 248 years (Fig. 1.4).

The discovery was made by the Canada-France-Hawaii telescope, during part of a study by the Outer Solar System Origins Survey (OSSOS). With hundreds of objects already identified and logged, perhaps other discoveries of bodies of a similar size to 2015 RR245 will follow, reinforcing the decision that Pluto, despite its diameter of 2300 km, should not be classed as a planet. There is some variance in the estimation of the size of Pluto, with the top-end diameter placing Pluto, the reclassified dwarf planet, at 2400 km.

This means that its declassification makes Pluto just another dwarf planet in the Kuiper Belt region, along with Eris, Haumea, and Makemake, although not all of these objects conform to

Fig. 1.4 Pluto, as captured by NASA's New Horizons spacecraft. The image is a high-resolution enhanced color view, July 14, 2015 (courtesy of NASA)

Pluto's 'planetary' shape, which makes for an uncomfortable tag being applied to all four.

Haumea, named after the Hawaiian goddess of childbirth, was discovered by Michael E. Brown, Chad Trujillo and David Rabinowitz of Caltech at the Palomar Observatory near San Diego, California, on December 28, 2004. The discovery was also claimed independently by a team headed by José Ortiz Moreno at the Sierra Nevada Observatory in Spain. However, despite being the first to file a claim to the Minor Planet Center in July 2005, the claim has been disputed.

Haumea's shape would tend to categorize the body more in line with that of an asteroid, elongated in nature, with its longest dimension measuring about 1960 km. Its shape classifies Maumea as a triaxial ellipsoid, with its major axis twice as long as its minor axis. The extreme elongation is probably caused by its rapid rotation, and by the probability that Haumea is the result of a collision. A day on Haumea lasts just under four hours, and the body is rated with a high albedo (brightness), thanks to its surface of crystalline water-ice.

As the largest member of the "collisional family," it has been suggested that Haumea and various other objects collectively are the dispersed fragments that resulted from a giant impact, with Haumea managing to hold onto two other lumps of debris—moons, if you will—that orbit the dwarf planet. This collisional family also includes some large trans-Neptunian objects (TNOs).

As Haumea was discovered during the Christmas period, the Caltech discovery team unofficially used the nickname "Santa" for the dwarf planet, with the two designated moons, named Hi'iaka and Namaka, being referred to at the time as "Rudolph" and "Blitzen".

Michael Brown and his team also discovered Eris, on January 5, 2005, with an estimated surface temperature of −231 °C. Eris was once considered to be the tenth planet, orbiting beyond that of Pluto, with its furthest point away from the Sun (its aphelion) actually taking it outside the Kuiper Belt. By the same token, at its closest point to the Sun (its perihelion), it is closer than Pluto's most distant point. Its orbital period is calculated at 560.9 years, compared with 248 years for Pluto. Eris was once potentially destined to join Pluto as a planetary body, but the reverse occurred, with Pluto being declassified and downgraded to match the status of Eris.

The diameter of Eris is slightly larger than that of Pluto, taking the lower size of 2299 km as Pluto's width. At 2325 km, this makes Eris the largest of the dwarf planets, although there continues to be some debate, given that Pluto's measurements are almost identical. Hence, there have been some references to Eris being Pluto's twin—a similar approach to viewing Earth and Venus sister planets.

With Xena, Lila, and Persephone (Pluto's wife in Greek mythology) all rejected as a name, Eris was chosen, from the Greek goddess of discord. However, like Pluto, the International Astronomical Union deems Eris to be a dwarf planet, so—with Pluto's demotion sealing Eris' classification—the existence of a tenth planet wasn't to be. Eris also has a moon of its own: Dysnomia.

Makemake was also discovered by Brown and his team on March 31, 2005, the findings being announced on July 29, 2005. Measuring 1400 km in diameter, Makemake takes around 310 years to orbit the Sun, with a surface temperature of −239 °C.

As the second furthest dwarf planet from the Sun, and the third largest dwarf planet in the Solar System, Makemake's surface is of a high albedo, making it the second brightest Kuiper Belt object after Pluto. Makemake was named after the creator of humanity, and god of fertility, in the mythos of the Rapa Nui (the native people of Easter Island). The name was partly chosen because its discovery date was close to Easter.

In April 2016, the Hubble Space Telescope was able to discern that Makemake had a companion, about 160 km in diameter, orbiting at a distance around 20,100 km. Unlike Pluto, Makemake's orbit is sufficiently distant from Neptune to allow for significantly less influence from Neptune on its path around the Sun, that distance placing it in a reasonably stable trajectory.

Debris for Thought

The category of cosmic debris has an order of its own, and probably offers up the most intriguing and taxing questions about the universe that we still grapple with. For throughout the universe, debris shape and size are manyfold with an array of guises. Collectively like the moving parts of a machine, their integration and ultimate cause and effect is greater than the sum of all the debris parts.

Within the seemingly disordered realm of these 'free radicals,' an order does exist. It is an order that, when encompassed into the mechanics of cosmology, plays just as important a part in our daily lives as the everyday observable workings of the universe, from the rising and setting of the Sun to the rising and setting of the Moon.

One of the biggest mistakes that humankind could make would be to not include the tiniest fragment of our known knowledge when considering the universe as a whole, for here lies the very nature of our own existence, perhaps even the very start of life itself—the foundations, the building blocks, from which the mighty rise and fall.

As we travel great distances in a year, not just orbital on the Earth's axis or around the Sun, cosmic debris travels with us, often unseen but ever present, with the ability to perhaps create, but most definitely to destroy.

One must not forget the powerful argument of an oscillating universe, and with it the belief that everything has always been here, and that perhaps humankind has merely joined the party at one particular instance during this ongoing venture. Quite possibly, Earth and its existence in the scheme of things is merely a passing phase within the life of the universe, which perhaps other 'Earths' have already experienced.

It may be a daunting prospect and to the naïve a reality check but, nevertheless, like all things, our own journey in this universe will end. Although the stony boulder deep in space knows no calendar, ours is marked even before perhaps the intervention of a cataclysmic strike, with the 'by the book' end courtesy of our own life-giving Sun expanding to fry Earth in the Sun's very own final exhalation.

The most likely end before our Sun gulps its last breath is the second 'end scenario' by the very event that may well have created us—a fitting end, some would say, with the very giver of life being the one that finally takes it away, rather like a massive cosmic boomerang finally completing its inward journey to its original thrower many, many, moons ago.

This is why the remnants of our universe play more than just a passing cameo: they are our past, our present, and our future. From every wandering chunk of rock that is captured by the gravitational pull of a planet and 'brought into order' to every pebble that burns up in our atmosphere, they all count; every single one of them has a place in the order of our universe, and every single one has the ability to cause a knock-on effect that could change everything.

Cosmic debris, while bunched together, is a kind of classless classification: it does not conform, behave, or sit nicely in any one neat folder. To think that there are collectively millions of objects roaming free throughout the universe is both frightening and awe-inspiring. For here, in our very backyard, lie the mavericks, the non-conformists, the envoys of creation, with a whole vista of possibilities laid out before them. But wait a minute; if they don't conform or slot in—if this rogue element just simply does not compute—then surely the remaining larger percentage of the universe's mechanics cannot possibly function? That, though, is the entire point. Their randomness and behavior is but a necessary part of the entire sum, an order within the chaos. This is the ultimate counterbalance within the universe, the random element.

Silently meandering, where little light breaks the darkness in the cold and icy depths of deep space, these wandering rocks are real pioneers from when the clock starting ticking, the last bastions of true astronomical free thought. They are the moving beacons among the stars from a time we can only picture in our imagination, where perception of what it was like before the Big Bang is personalized and tailored to each individual, a private visualization of an echo from creation itself. These rocks are the dismembered foundation of all that formulates the way in which we each lead our lives, the cornerstone on which we base our existence.

Down Comes the Rain

Researchers have found evidence to suggest that Earth has been or is still being bombarded by ash particles from a supernova. It is not quite rain as we know it, but a fine sprinkling of cosmic debris.

The debris sports a rare isotope, one that only appears in the aftermath of a star collapsing. The detection, made by the Cosmic Ray Isotope Spectrometer (CRIS), aboard NASA's Advanced Composition Explorer, is so rare that only a handful have been detected.

Therefore, we are literally being rained on by stardust. From within the stars we came, to the stars we shall go.

2. Minor Planets and Asteroids

A Curious Grouping

The term 'minor planet' is an astronomical classification that encompasses minor planets and asteroids. However, to arrive at this classification, the understanding of where other bodies fit in is necessary. The International Astronomical Union defines a planet as a body in orbit around a star, which has the ability to pull its mass into a rounded shape. The path on which the planet travels has also been cleared of smaller bodies, so that the planet has total dominance.

Disregarding any natural satellite that a planet may have, a dwarf planet has the characteristics of a planet but does not possess the size to have swept its orbit of other debris. Dwarf planets include Pluto (declassified from being a fully-fledged planet), Ceres, Eris, Haumea and Makemake. The term 'minor planet' can encompass not only dwarf planets but also asteroids, Trojans, centaurs, Kuiper Belt objects, and trans-Neptunian objects—all of which are discussed in more detail in the coming chapters. With the orbits of 700,000 of these minor planets already recorded by the Minor Planet Center, the cataloging of such objects is ongoing, and with the potential for reclassification always a possibility, it would seem to be a lengthy task that may never be completed.

Although asteroids fall into the category of minor planets, or planetoids, as some scientists suggest they should be termed, there remain others who are at odds with classifying asteroids alongside such a great diversity of other bodies. There would seem to be justification to the objection, with the Asteroid Belt—where so many such bodies collectively lie—a major part of the argument.

Trojans are known and cataloged objects that orbit the Sun, subdivided into the vicinity in which they circuit our star. The first grouping is Earth Trojans, bodies that share the same space as Earth shares the space with the Sun, orbiting in the same fashion, and displaying a constant regularity in motion as Earth does.

© Springer International Publishing AG 2017
J. Powell, *Cosmic Debris*, Astronomers' Universe,
DOI 10.1007/978-3-319-51016-3_2

One such Earth Trojan, which does not add to a substantial cataloged list by any means, is 2010 TK, measuring 300 m in diameter and discovered by WISE, the Wide-field Infrared Survey Explorer.

Named after characters in Homer's *Iliad*, there are specific guidelines to make a Trojan body what it is, and it is these guidelines that consequently discount several other known bodies orbiting Earth that don't quite fall into the same category as 2010 TK.

The criteria for the Trojan relates to Lagrangian points. Lagrangian points are any positions in an orbital configuration of two large bodies (in the case of 2010 TK, the Sun and Earth) where such a body, affected only by gravity, can maintain a stable position. These points mark the position where the Sun and Earth offer their greatest gravitational pull on 2010 TK, providing the exact force to allow this body to orbit with them. There are five Lagrangian points (L1 to L5), named after an essay published in 1772 by Italian mathematician and astronomer Joseph-Louis Lagrange (1736–1813), entitled "Essay on the Three-Body Problem".

The Asteroids

The common perception of an asteroid is not a favorable one, despite it also having the potential to deliver life as well as take it away. The general assembly of asteroids within the Asteroid Belt pose little threat if order and conformity reigns, but as we will discover later in the book, it appears to be the intervention of a 'foreign' body into the order of the belt that causes the equilibrium to be disturbed.

Our asteroid journey starts with our understanding of them—what they are, how they work, and what place they have in our universe. It is also necessary to take away the idea that asteroids are merely a nuisance, bodies that have to be sought out and identified. Their presence in the universe is both necessary and important, for balance if nothing else.

The vast majority of these metallic, rocky bodies orbit the Sun in the Asteroid Belt positioned between Mars and Jupiter.

There are a number of theories as to how the Asteroid Belt came into existence, and the absolute conclusion remains a matter of much debate.

It is probably the vast and extensive gravitational field of Jupiter that prevented the contents of the belt from forming into a planet-sized body, with the 'binding' motion of material being continually disturbed, preventing any long-term melding. Given that the total estimated mass of all the asteroids combined would present an object of around 1497 km in diameter that, in itself, would cause a problem of classification, as the body would be less than half the size of our own Moon. However, the formation would have taken place long before anyone even considered categorizing them, so it's all pretty irrelevant.

Whether pulled apart prior to its complete formation by external forces, or possibly destroyed before evolving sufficiently far, could this planet possibly have existed in the formation of the Solar System? It seems unlikely. The more accepted theory is that the asteroids in the belt are not those of a planet that never formed but the debris left by the formation of other planets. The Asteroid Belt may well just be a dumping ground for a substantial amount of leftover material from the early evolution of the Solar System, where, with gravitational forces at work, debris of all sizes became shepherded into an orbit, with the debris itself becoming the gravitational attraction over time for yet more rubble.

These airless rocks would appear to be nicely contained within the parameters of the belt, but all is not constant within the asteroid community, with plenty of jostling, and of course always the prospect of the intervention from a foreign body in the established domain. Such an intervention would have a knock-on effect of destabilizing an area, with perhaps far-reaching consequences.

Compared to the planets in our Solar System, although the Asteroid Belt orbit has been likened to a doughnut in shape, the asteroids generally match the average orbital eccentricity of the major planets. The bulk of the belt's contents are inclined at 20 or 30 degrees to Earth's orbit, with a smaller proportion holding a more radical path around the Sun.

The belt is not an overly congested place; in fact, the distribution of asteroids within it makes for a rather vacant appearance, given the volume of space through which they have to travel. It has been calculated that the average distance between objects is 965,606 km, therefore making the 'gap' between asteroids some 24 Earth circumferences.

Despite our mind's eye visualization offering us this vast area of tumbling rocks, varying in proportion and size, it is in fact comprised mostly of empty space. Astronomers know of the existence of clumping in the belt, where some asteroids do travel together.

In 1866, astronomer Daniel Kirkwood (1814–1895) noticed that certain gaps or dips in the distribution of asteroids corresponded with the locations of orbital resonances with Jupiter. It was almost as if asteroids had been pulled out of position at these "meeting points," leaving significant gaps in the belt, completely void of any bodies.

Kirkwood explained the disappearance of the asteroids with the regular interventions by Jupiter during its orbital period. In essence, whenever Jupiter approached, whatever lay in the belt would be drawn or pushed clear, creating a Kirkwood gap. There is significant research to suggest that these resonances may well force debris from the Asteroid Belt towards Earth, or, at least in part, offer an explanation for the realignment of many an asteroid's orbital path.

As for the actual amount of asteroids contained not just in the belt but in the entire Solar System, this again is a figure continually on the rise as observational methods become more refined. At the present time, 150 million seems a conservative but plausible estimation.

Ceres—King of the Asteroids

Within the belt a vast array of sizes and shapes exist, the largest of which is Ceres. The vast majority of the bodies follow a stable, slightly elliptical orbit, revolving in the same direction as the Earth, with a full circuit of the Sun taking from three to six years (Fig. 2.1).

Ceres, Vesta, Pallas, and Hygiea account for about half the total mass of the Asteroid Belt. Ceres measures 945 km in

Fig. 2.1 Artist's impression of NASA's Dawn mission to Ceres (courtesy of NASA)

diameter, with Vesta measuring 525 km in diameter and Pallas a close disputed third, at 512 km in diameter. Hygiea measures 499 km in diameter with its oblong shape affording it a range of size from 349 to 499 km.

Positioned 299 to 595 km away from the Sun, it is, as you would expect, a cold existence for the asteroids in the belt. Tumbling and spinning, temperatures on these worlds can vary greatly from whichever side is facing first solar radiation to the vast depths of the stellar background faced by the other side. A typical average temperature on an asteroid would therefore range between −73 °C and −108 °C.

Ceres was discovered by Sicilian monk and astronomer Giuseppe Piazzi (1746–1826) at Palermo, Sicily, on January 1, 1890. It was discovered after Piazzi had followed up on mathematical calculations that indicated that there was a planet positioned in between the orbits of Mars and Jupiter. Piazzi named his discovery after the Roman goddess of harvests and corn.

Despite the calculations being somewhat awry, Ceres was spotted, and originally classed as a planet until being reclassified in the 1850s as an asteroid, after subsequent objects of a similar nature were discovered. Piazzi's work and discovery also at the time undermined research that was being conducted at the time by a group of scientists calling themselves the "Celestial Police."

Working as a team and with the use of the then prominent observatory owned by German astronomer Johann Hieronymus Schröter (1745–1816) at Lilienthal in Bremen, the "Himmelspolizei" were so-called for the ambitious project of wanting to bring order to the Solar System, through the discovery and subsequent classification of bodies. The group, which included the then current British astronomer royal, Rev. Dr. Nevil Maskelyne (1732–1811) and French astronomer Charles Messier (1730–1817), was completely driven by the prospect that another planet existed. With an allocated 15-degree arc allocated to each member to locate this postulated body, they worked tirelessly in its pursuit, only to be thwarted in the quest by lone observer Piazzi.

Observing from the Palermo Observatory, Piazzi originally believed that the point of light he had spotted in the night sky might be a comet. Over the coming weeks, he began to think differently, as the initial finding, first sighted in the constellation of Taurus, did not conform to that of a comet. For one thing, its appearance was not fuzzy or diffused and, for another, its slow and uniform movement did not agree with that of a comet against the stars.

In the early seventeenth century, German astronomer Johannes Kepler (1571–1830) had speculated that somewhere between the orbits of Mars and Jupiter a planet must exist. Kepler, discoverer of the basic laws of planetary movement, along with his countryman Johann Titius (1729–1996), showed that simple mathematical calculations produced a numerical formula giving the conclusion that, given the spacing of then known planets from our Sun, it was a natural matter of order that a planet should exist in an orbit beyond that of Mars and before the orbit of gas giant Jupiter. Granted, there were a number of striking anomalies in the formula, but basically it held true. However, despite these apparent flaws, the formula did predict the location of the planet Uranus in 1781.

However, it seems that coincidence rather than mathematics governed its working, as the numbers simply didn't add up following the discovery of Neptune in 1846 and later Pluto, in 1930. With chance seeming to play more of a part, the formula rapidly lost credibility over time. Therefore, the discovery of Ceres—

which caused initial delight at the apparent confirmation that the formula did work in the predicting of a planet in between the orbits of Mars and Jupiter—quickly turned to puzzlement, as other bodies in the then unknown Asteroid Belt were found.

Between 1801 and 1808, astronomers tracked down a further three asteroids in the region of space where Ceres had been located, namely Pallas, Juno, and Vesta, each smaller than Ceres. A new explanation was now required to account for the presence of these bodies, and for the probability that these four were not alone. This new finding acted as a catalyst in the astronomical community to find answers, and it was deemed that perhaps a new class of celestial object should be created. As further years passed and new discoveries were indeed made, it became clear that there appeared to be an entire belt of these objects orbiting the Sun.

However, following the success of finding Pallas, Juno, and Vesta in relatively quick succession, there was a distinct gap in new findings of any object in that region of space, so much so that many astronomers, having initially been whipped up in the enthusiasm to find more asteroids, declared that there were simply no more to be found, and turned to other studies. For a time, the search that had initially captured the world of science was readily abandoned by most.

However, German amateur astronomer Karl Ludwig Hencke (1793–1866) was to pick up the thread of searching for asteroids in 1830, and after 15 years of scouring the skies discovered Astraea, the first new asteroid to be found in 38 years. Less than two years later he struck gold again, with the discovery of Hebe. Upon seeing the discoveries made by Hencke, many other astronomers now turned their own 'scopes away from current projects back to prospecting for asteroids, believing that more were yet undiscovered. In the years that followed, the discovery rate was to average out as a find a year, with the exception of 1945.

Hencke's initial new finding also acted as a catalyst in the astronomical community to find answers, and it was deemed that perhaps a new class of celestial object should be created. In 1891, astrophotography was the new tool in the hunt, with German astronomer Max Wolf (1863–1932) pioneering its use to detect further bodies, which appeared as short streaks on long-exposure photographic plates. Wolf went onto find an incredible 248 asteroids using this method, going somewhere near to doubling the amount of those already cataloged by using standard search patterns.

With the rate of discoveries at such a high, an ironic twist in the hunt occurred. From the painstaking searches of yesteryear—where calculations followed by more calculations and then ultimate triumph and disappointment were endured—Max Wolf's revolutionary method of seeking out asteroids had turned searching on its head. An era had now been entered into wherein it was generally considered that there must be literally hundreds of these objects orbiting between Mars and Jupiter, and the pursuit became akin to shooting fish in a barrel—easy pickings, if you will, and not worthy of pursuit. So great was the disdain towards the prospect of just finding yet another asteroid that astronomers now simply referred to asteroids as "vermin of the skies," a phrase variously attributed to Eduard Suess (1831–1914) and Edmund Weiss (1837–1917).

Types of Asteroids

The majority of asteroids fit into three distinct groups, although there are other, lesser groups. The first is C-type (carbonaceous), which accounts for more than 75% of all known bodies. These asteroids are very dark in nature, more red in hue than other bodies, offering little in the way of reflectivity. Reflectivity is measured in terms of albedo, or intrinsic brightness. A body that reflects light with a perfect surface for doing so has an albedo of 1.0. A body whose surface is black—a surface perfect for absorbing light and returning no reflective nature—is classed as 0.0. The C-type asteroids fall within an albedo bracket of 0.03–0.09, generally positioning themselves to the outer regions of the belt. Chemically, their spectra match the primordial composition of the early Solar System, with the omission of only the lighter elements removed.

The second most predominant class of asteroid is S-type (silicaceous, silicate-rich), which accounts for 17% of all known bodies. S-type asteroids are relatively bright in nature, with an albedo of 0.10–0.22. Positioned and tending to dominate the inner portion of the Asteroid Belt, they are metallic in nature, with iron and magnesium silicates. No significant carbonaceous compounds are to be found within them. The overall composition of the S-type indicates that their materials have been significantly modified from their primordial composition, probably through a process of melting and reformation.

The third tier of classification makes up the rest of the belt's bodies, M-type (metallic, metal-rich). Their composition is generally dominated by the presence of iron. Dwelling in the middle region of the asteroid belt, they are, like the S-type asteroid, classed as relatively bright, 0.10–0.18. Although there remains some doubt as to whether all M-type asteroids are compositionally similar, a proportion is believed to have formed from the metallic cores of other bodies whose components were disrupted through collision. Therefore the M-type category could be considered a bit of a dumping ground classification for all asteroids that don't fit into either the C-type or S-type—a miscellaneous collection that generally accounts for the makeup of every other known asteroid. As with all classifications, if numbers of a certain type of asteroid within the M-type were to show a high percentage, this would undoubtedly force another class to come into being. This was to become apparent when subsequent discoveries were made, largely thanks to the plethora of spacecraft that were to make close-up studies of the Asteroid Belt and its contents.

The Galileo Spacecraft

We largely have three main sources to thank for the classification of asteroids. The most notable source has been the flybys of NASA's Galileo spacecraft, launched in October 1989, with the mission to make two close-up observations of asteroids 951 Gaspra and 243 Ida, before moving on to study Jupiter. The Galileo mission consisted of two spacecraft, an orbiter and an atmospheric probe (Fig. 2.2).

Costing NASA $1.39 billion with a further contribution of $110 million from international sources, the mission, which involved contributions from 800 people, seemed a far and distinct echo from the man it was named after, Galileo Galilei (1564–1642).

Fig. 2.2 A study of Europa, as part of NASA's Galileo mission to Jupiter (courtesy of NASA)

Often referred to as "the father of science" Galileo's contributions reach far beyond the time period in which he lived, with the NASA mission reflecting in some way a physical representation of just how far Galileo's mind could extend. This craft symbolized his insight into worlds beyond, as in 1610, when he spotted what we now know are the four moons of Jupiter.

Although Jupiter was the main target of the Galileo mission, astronomers with more of an interest in asteroids were to receive, during the course of the craft's epic journey, a fascinating insight into these worlds. With the largest planet in the Solar System as a target, the accuracy of Galileo's trajectory lay in the navigation to the much smaller bodies of our Solar System, a logistical execution requiring a high level of precision. Ultimately, though, these rendezvouses were also a great success.

Launched off the back the space shuttle Atlantis on mission STS-34, Galileo would cleverly use the gravitational fields of both Venus and Earth to build up enough velocity to propel it towards Jupiter. Observations were made during this inward part of the journey of Venus, Earth, and the Moon, before its outward trip towards the Asteroid Belt and then onto its primary target of Jupiter.

Galileo was the first spacecraft to orbit Jupiter and launch an entry probe into the planet's atmosphere. However, there was to be an unexpected and substantial bonus to the mission. This came in the form of the infamous strike on Jupiter by Comet Shoemaker-Levy, with NASA scientists afforded front row seats to witness the fragments of the comet as they ploughed into Jupiter's clouds. How fortuitous that a probe just happened to be in the vicinity at the time, there on hand, to capture the remarkable spectacle that was to unfold.

However, prior to this landmark astronomical surprise, back in 1991 after two months plowing through the Asteroid Belt, Galileo notched up another astronomical first, rendezvousing at close range with an asteroid.

Passing just 1593 km from 951 Gaspra, Galileo was able to photograph, take measurements, and assess the composition and physical properties of the asteroid. The photographs revealed a cratered, very irregular body, measuring roughly 19 by 11 km.

In 1993, Galileo made its second rendezvous, this time flying within 2414 km of asteroid 243 Ida, which gave scientists yet another first. This particular revelation was the discovery that 243 Ida was not alone in the vastness of space, appearing to be accompanied by its own moon, which was subsequently dubbed Dactyl, measuring just below 2 km in diameter. This in itself produced amazement and provided a further puzzle in the world of cosmic debris. Here we had an asteroid that had a moon! With measurements and analysis taken, little Dactyl was discovered to be spectrally different from 243 Ida, subsequently classed as an SII subtype S-type asteroid. In later years, more than 150 asteroids were shown to boast a moon for company and, in some cases, even two moons.

It also became apparent that the asteroid/moon relationship would not necessarily be governed by size. Some asteroids were discovered with almost equal size, making them a sort of binary system, orbiting each other, as they both orbit the Sun.

Galileo had also opened our eyes to yet another twist in the cosmic debris tale, with the spacecraft's journey making significant advances in our understanding of Jupiter and its four inner moons, Io, Europa, Ganymede, and Callisto. The latter three moons yield support for the theory of a liquid ocean under the icy surface of Europa, with indications of similar liquid saltwater layers under the surface of Ganymede and Callisto.

Like all missions, it was with a strange yet inevitable sadness that, having delivered so much, this extraordinary space journey would come to an end. After 14 years in space and having spent eight years studying the Jovian system, Galileo was to meet its end with a final dive into Jupiter's atmosphere at 48 km/s. Even as it descended into the Jovian clouds, this plucky craft was still able to transmit data. Its destruction, although honorable, was also necessary, as Galileo could not be left to roam the Jovian system unattended, as there was the possibility that one day the craft could have crash landed on one of Jupiter's moons, potentially contaminating it with terrestrial bacteria. Therefore, Galileo had to be destroyed.

The list of Galileo's achievements is remarkable, all despite problems that saw NASA technicians constantly battling to keep the craft functioning during its mission. One of the most serious difficulties encountered by NASA was Galileo's failure to deploy the main antenna, an issue that was corrected by rewriting the probes on-board software in order for Galileo to deploy its reserve antenna. If that had failed, the mission would have sunk without trace.

The diary of Galileo encounters started with the initial flyby of Venus in 1990. The craft would use Venus's gravity to gain the necessary velocity for the next phase of its trip—from Venus to 951 Gaspra in 1991, then on to the 243 Ida encounter in 1993. The intervening year, 1992, saw a flyby of Earth before the main rendezvous at Jupiter. Galileo's encounter with Jupiter coincided in July 1994 with observations of the 20 fragments of Comet Shoemaker-Levy as it plunged into Jupiter's night-side atmosphere over a six-day interval.

Galileo also managed to confirm a supposed but not substantiated feature on our own Moon. The craft discovered an ancient impact basin in the southern hemisphere of the far side of the Moon, something previous Apollo missions had inferred but had never had the chance to properly map. The spacecraft also discovered that the Moon probably had much more in the way of

Fig. 2.3 NASA's Cassini spacecraft took this photograph during the closest-ever flyby of Saturn's moon Mimas. The giant Herschel impact crater, measuring 246 km in diameter, dominates (courtesy of NASA)

lunar volcanism than previously thought, adding to the rich tapestry of not only impact craters on the Moon's surface but perhaps a more internal explanation for its outward appearance.

Galileo represents one of the many missions to have visited asteroids. Other notable expeditions have included a joint mission by NASA and the ESA in 1997 that, en route to Saturn, took the Cassini spacecraft directly through the Asteroid Belt; the European Space Agency's Comet Mission, which flew past the asteroids Steins and Lutetia in 2004; and NASA's Dawn mission, which encountered Ceres and Vesta in 2007 (Fig. 2.3).

The NEAR Mission

It was NASA's Near-Earth Asteroid Rendezvous (NEAR) mission that took visiting and photographing asteroids to a new level.

Towards the end of its time spent encountering 433 Eros, an S-type asteroid, the decision was taken to attempt to actually land on its surface.

Discovered by German astronomer Carl Gustav Witt (1866-1946) on August 13, 1898, 433 Eros is a near Earth asteroid (NEA), part of a group of NEAs known as Amor asteroids, named after 1221 Amor. Amor asteroids approach the orbit of Earth but do not intersect it, with the majority of these bodies tending to cross the orbit of Mars instead. It is worth noting that, although credit has been allocated to Witt for this discovery, it was in fact jointly discovered with observations noted on the same date by Auguste H. Charlois (1864–1910) in Nice, France.

Witt, director of the Urania Observatory in Berlin, discovered two asteroids, of which 433 Eros was the most notable, being the first NEA to be discovered and the first asteroid to be given a male name. Measuring 34 × 11 × 11 km, 433 Eros registered its closest approach to Earth on January 23, 1975, at a distance of 22 million km. However, bar a collision to alter its course, 433 Eros is not a threat to Earth. Witt's second discovery was 422 Berolina, found within the main Asteroid Belt.

Launched in 1996, the NEAR mission's epic journey to 433 Eros was to take 14 years, with February 2000 seeing the craft finally achieve orbit around the asteroid. NEAR and its whole mission program had not been designed for such an eventuality but, late into the mission, NEAR Shoemaker, to give the probe its full title (Shoemaker being in honor of planetary scientist Gene Shoemaker), made a daring and unscheduled descent onto the surface of the asteroid.

During its final 4–5 km before landing, NEAR was able to take dozens of detailed pictures of its soon to be host, transmitting back to Earth the most comprehensive photographic images ever obtained of an asteroid. With less than 400 feet to go, objects down to the size of golfballs were photographed, with NEAR finally touching down on 433 Eros near a large depression on the surface named Himeros. With the final stages to touchdown speed at just 4 mph, NEAR made a gentle landing on the asteroid's surface and, given the terrain, a most fortunate one.

NEAR, now operating outside of its stated mission, continued to transmit data from the surface of 433 Eros, sending back priceless information about the composition and chemistry of this small world. It was a truly remarkable feat, with data continually being sent until it finally ceased operating in February 2001.

NASA's mission statement for NEAR included that the probe would make the first-ever attempt to orbit an asteroid, with other objectives including examining 433 Eros' magnetic field, its interaction with the solar wind and, ultimately, to link 433 Eros to meteorites that had been recovered on Earth.

If NEAR were to collect sufficient information on chemical composition, then a causal link could be established between 433 Eros and other S-type asteroids and, more importantly, with meteorite pieces of S-type that had been found on Earth—perhaps even linked to 433 Eros itself.

In essence, NEAR's mission was to attempt to answer by its discoveries some of the fundamental questions about the origin of asteroids that were close to Earth's orbit. In turn, these objects could contain pointers and clues about the formation of Earth and other planets in the Solar System. 433 Eros itself and the various possibilities made it a tempting proposition for NASA, for here we had an asteroid whose surface looked in pristine condition, perhaps offering itself up as a relic from a time in space, some 4.5 billion years ago, when conditions had evolved to the point when our very own Earth came into existence.

The data collected from NEAR was groundbreaking as, from it, a significant and thorough profile of an asteroid was to be established, the like of which had never been documented before. 433 Eros presented itself as a solid, undifferentiated, primitive body, a distant throwback to the creation of our Solar System. The chemical composition analysis allowed scientists to determine that 433 Eros was not just a "rubble pile" but, rather, a consolidated object. In turn, this would enhance our understanding of the relationship between asteroids and meteorite debris found on Earth. Littered with impact craters from its turbulent past—over 100,000 discerned from the 160,000 images taken by NEAR—this asteroid proved to be a museum piece from time, a 'hands-on' chunk of history.

Near Earth Asteroids

The relationship between asteroids and meteorites has been continually unraveling over time as more and more is discovered about them. Ultimately, however, the link does remain a mystery, a very gray and somewhat annoying puzzle.

The most common meteorite, the ordinary chondrite, is composed of small grains of rock, with the overall appearance of a chondrite being unchanged since the creation of our Solar System. Other types of meteorites discovered, being stony-iron, are probably the remnants from the melting of much larger bodies, so that the heavier metals and lighter rocks subsequently separated into different layers. Considering that S-type asteroids are the most common of known asteroids, could there be a link between these asteroids and the abundance of ordinary chondrites? Spectral analysis suggests that S-type asteroids may be geochemically processed bodies, as opposed to meteorites of a stony-iron composition. If S-types are unrelated to ordinary chondrites, then what are they related to? By the same token, if there is a link between S-type asteroids and ordinary chondrites, an explanation for this spectral disagreement must be found.

Aside from the main Asteroid Belt, which harbors the vast majority of bodies, another classification of asteroid (not involving their composition but their movement in relation to Earth) is that of NEA's or near Earth asteroids. As mentioned with 433 Eros, these bodies don't conform to the relatively stable and consistent orbit offered to us from within the belt but travel their own path in the Solar System and are so-called because their orbits bring them within 190 million km of Earth.

These more maverick bodies may well be the result of knocks and bumps from other asteroids in the main belt, with perhaps the body eventually sprung from the belt by the influence of Jupiter, sending them into a different orbital plane around the Sun and, on occasion, a deflected passage that takes them near to Earth. The analogy of pinball, billiards or snooker within the Asteroid Belt has a serious edge to it, for it is here during, perhaps, a weighty engagement between two bodies that an asteroid could be directed in our path.

Apart from the aforementioned Amor asteroids, there are several other NEA groupings of note. The first of these are the Apollo asteroids, which dissect Earth's orbit over a period greater than one year, and the Atens, which cross Earth's orbit in a period of less than one year.

Estimates vary as to how many NEAs exist, but what has been discovered probably only accounts for a small proportion of what is actually out there. NEAs are essentially a young population, with their orbits having been tempered and tailored over a 100-million-year timescale, following possible collisions inside and outside of the Asteroid Belt, and the influence from the larger outer planets. Within the ever-changing estimates, it is thought that at least a thousand NEAs exist outside of a 0.5–1 km sweep of our orbit and, given their size and speed, are subsequently considered to pose a threat to Earth.

On average, three new NEAs are discovered daily, with discoveries by the end of 2015 totaling 13,024. A total of 1609 of these are classed as potentially hazardous asteroids, (PHAs). All have the ability to cause destruction, but, as of yet, none of these has a specified orbit in the future that will bring them into contact with Earth. There may be near misses in astronomical terms, but no direct hit.

Phobos and Deimos

So what of other asteroids that aren't either corralled into the Asteroid Belt, or classed as NEAs? Could a small proportion of asteroids be orbiting the planets either side of the belt? In particular, the Martian moons, Phobos and Deimos?

Although it is likely that Jupiter has caught a fair share of asteroids, largely small in nature and generally situated in the outer reaches of the 70-plus known Jovian system, the debate over how the Martian moons came to be remains.

For many years it was considered that Mars had no moon. German mathematician, astronomer, and astrologer Johannes Kepler suggested the possibility of the existence of bodies orbiting the planet (Fig. 2.4).

Fig. 2.4 The Martian moons, Phobos (nearest) and Deimos (courtesy of NASA)

The same possibility was referred to by Jonathan Swift (1667–1745) in his satire *Gulliver's Travels* (1726), with speculation on what was to become a truth, possibly founded in the common belief that if Earth had one moon, and Jupiter at the time was known to have four moons, surely Mars must also have a moon or moons, given that Mercury or Venus didn't have any at all.

Whatever the reasoning, Swift, an essayist, poet and cleric, produced substantial calculations to reinforce his belief that Martian moons existed, his later work probably combined with input from his close friend John Arbuthnot (1667–1735), a physician and mathematician.

In honor of the contributions of Swift and of writer Francois-Marie Arouet (1694–1778), who wrote under the non de plume Voltaire, two craters on Deimos were subsequently named Swift and Voltaire. In his short story of 1750 entitled "Micromegas," Voltaire had predicted that Mars had two moons.

American astronomer Asaph Hall (1829–1907), with the use of a 26-inch refractor, made the discovery of Deimos on August 12, 1877, followed six days later by that of Phobos. At the time, Hall was actively searching for moons around Mars, with an earlier sighting that he had made being thwarted by bad weather, which meant he was unable to confirm what he had seen. No matter, his persistence paid off after this setback on August 10 as, two days later, he was able to confirm the existence of Deimos.

Phobos, meaning panic or fear, and Deimos, meaning terror or dread, were named after the horses that pulled the chariot of the Greek war god Ares, known as Mars to the Romans. However, according to other sources, the two moons are named after the two sons of Ares, who attended their father in battle.

Phobos, measures just below 23 km in diameter, with an orbital period of 7.66 h. Deimos is slightly smaller at just under 13 km, with a far greater but still rapid orbit of 30.35 h.

It is known that the orbits of the two moons aren't stable, and that Phobos is making a slow but steady descent towards the Martian surface, some 6 feet every century. This is not an immediate issue, but this spiraling orbit will see Phobos eventually destroyed on the planet's surface or, perhaps before that fate, torn into a subsequent ring of rubble around Mars. As for Deimos, the second of the two moons is showing signs of drifting away from Mars. The moon seems set in its orbit, with the only possible alteration coming from outside influence, or following the demise of Deimos in some way. Phobos orbits at a distance of 5955 km from the Martian surface, while Deimos is much further away, at 20,069 km.

The speculation that Phobos and Deimos are captured asteroids is well founded, as they bear more of a resemblance to asteroids than they do to our own Moon. To start with, both are small, classing them as some of the smallest known moons in the Solar System. Secondly, both are made up of material akin to carbonaceous chondrites. Thirdly, their shape is elongated, as opposed to the more accepted and rounded moon-like shape. All the evidence seems to suggest that these moons are in fact asteroids that have perhaps been nudged by the gravitational field of Jupiter into the path of Mars, and subsequently, over time, came within the gravitational pull of the Red Planet and eventually into an orbit. Although Mars seems an unlikely candidate to be making such a capture, this doesn't mean to say that in the past it wasn't capable of influencing bodies to a much greater extent. Indeed, were the pull of Mars substantially greater in the past, the planet may not have needed Jupiter to nudge bodies in its direction, being capable of capturing its own unassisted.

The orbits of both Phobos and Deimos are not considered to be erratic ones; in fact they are near circular in nature. So if these are indeed captured bodies, surely they would have a more unstable, eccentric orbit, wouldn't they? If anything, their path mirrors that of Earth's own Moon. The 'normalizing' of the orbits of the Martian moons, which brings their paths into line with the orbital plane of Mars, could be explained by atmospheric drag and, considering time-frames involved, by possible tidal forces.

The captured asteroid theory still holds much sway, but this remains strongly challenged by two other theories. One is that they are fragments left over from the creation of Mars, with gravity perhaps shaping the fragments into these two oddly shaped rocks. The other is that the moons are the product of a collision of some sort near to the planet. Another avenue of speculation promotes the possibility of Phobos and Deimos once being a binary asteroid, torn apart from each other under the influence of Martian tidal forces. Is it possible that once the Martian skies were full of debris, orbiting the planet in belt format? If so, what happened to the belt, and why are only two of the original abundance of rocks remaining?

Both moons are tidally locked into their orbits, much like Earth's moon, always presenting the same face towards Mars. The reason for Phobos' decay in orbit is thought to be tidal, and this tidal pull may have accounted for earlier, smaller moons that possibly orbited Mars, the craters of which are littered on the planet's surface. In certain areas, strings of craters have been identified—generally speaking, the further from the equatorial region, the older (Fig. 2.5).

Jupiter Trojans and the Hilda Group of Asteroids

Aside from the Jovian and Martian moons, there are other 'colonies' that exist outside of the asteroid belt, notably Jupiter's Trojans and the Hilda group of asteroids. Trojan asteroids are bodies that revolve around the Sun in the same orbit as a planet, occupying stable positions, Lagrangian points. These points are fixed to the planet's position, lying either 60° ahead or 60° behind.

Fig. 2.5 Martian concretions found near the Fram Crater on Mars, which was visited by the Mars Exploration Rover Opportunity during April 2004. The area shown is 1.2 inches (3 cm) across (courtesy of NASA)

Along with the Sun, Jupiter is one of the two largest objects in the Solar System, and although Jupiter orbits the Sun as the other seven planets do, its gravitational pull is very strong, so much so that the planet actually pulls back at the Sun, almost canceling it out. It is here that these Lagrangian points are created, allowing other smaller objects—in the case of Jupiter, the Jovian Trojans—to travel along in Jupiter's orbit without being pulled out of position.

In 1990 an asteroid occupying similar Lagrangian points was discovered around the orbit of Mars. Named Eureka, this particular body has been joined by several other asteroids, collectively comprising the Martian Trojans. Since 2001, Neptune has been found to have its own Trojan. Despite finding other Trojans in the orbit of other planets, the name Trojan itself generally refers to the asteroids that accompany Jupiter.

Of the two groups of Jovian Trojans, there are in excess of 5000 known bodies, about 65% of them belonging to the lead group ahead of the planet with a cluster of 35% trailing behind the planet in the second grouping. The lead group ahead of Jupiter is known as the Greek Camp, with the Trojan Camp following. The first of Jupiter's Trojan asteroids was discovered by Max Wolf on February 22, 1906. It measures 135 km in diameter and was duly named 588 Achilles.

The term "Trojan asteroid" was coined when a decision had been taken to name all of the asteroids found in both camps after warriors in the Trojan War, Greek and Trojan, respectively. There are exceptions to the camp members, with Hector (a Trojan spy in the Greek camp) and Patroclus (a Greek spy in the Trojan camp), the first two Trojan asteroids to be discovered and named before the two camps were established.

The Jupiter Trojans are fairly elongated in their groupings, despite being locked into orbit with the planet. Saturn's gravitational field has been known to have an effect on these Trojans, causing them to oscillate. Although Saturn itself has no associated Trojans on its orbital plane, it is thought their absence is due to the greater gravitational pull of Jupiter, probably removing any such Lagrangian points from Saturn's orbit.

On a separate but equally stable orbit in relation to Jupiter is the Hilda group of asteroids. Sometimes referred to as the Hildian asteroid group, two collisional families exist within the collective: the Hilda family and the Schubart family. The "collisional" tag is given to an assembly of bodies that are thought to have a common collisional origin.

The founding member of the Hilda family is 153 Hilda, discovered by Austrian astronomer Johann Palisa (1848–1925) in 1875, and named after one of his daughters. Palisa was a prolific discoverer of asteroids, finding 122 such bodies.

The Schubart family is the second collisional family. The founding member, 1911 Schubart, was discovered on October 25, 1973, by Swiss astronomer Paul Wild (1925–2014) while working at the Zimmerwald Observatory in Berne, Switzerland. The finding was named in honor of German astronomer Joachim Schubart (b. 1928), who developed a technique for observing the motions of minor planets.

There are in excess of a thousand known Hilda group asteroids, many of which remain unnumbered. The Hildas are not considered to be the same as the Jupiter Trojans, since not all the members are physically related, with the only common ground being their relative closeness and shared orbit.

Although the Jupiter Trojans orbit the Sun with the same period as Jupiter, 11.9 years, the Hilda group takes several years less to make a circuit, 7.9 years, meaning that Jupiter orbits the Sun twice for every three orbits completed by the Hilda group.

Lagrangian points make for a stable 3:2 resonance pattern with the Hilda group, the asteroids confined to a basic triangle shape by the combined pull of Jupiter and the Sun. Jupiter always lies along one of the 'sides' of the triangle, so that each Hilda passes the line between the Sun and Jupiter when the body is near perihelion. In turn, this means the distance between the Hilda group and Jupiter remains the same.

Centaurs and the Kuiper Belt

There are a number of bodies that make up the remainder of asteroid-type objects in our Solar System, although classification does not view some of them as being 'outright' asteroids. The Centaurs are classed as such. In mythology, a centaur is half human, half horse. In astronomy, these mainly icy planetesimals are half asteroid, half comet.

The bulk of Centaurs orbit between Saturn and Uranus, with a lesser number orbiting between Jupiter and Neptune. Centaurs frequently cross the orbits of the outer planets, with gravitational interactions between the planets and the Centaurs, throwing many into very unstable orbits—orbits that are subsequently redefined again by further encounters.

The erratic nature of Centaurs has led some scientists to believe that these bodies could be comets in a very early stage of formation, proto-comets, with their origin in the Kuiper Belt. The Kuiper Belt (also known as the Edgeworth-Kuiper Belt) is a region of the Solar System that exists beyond Neptune, similar to the Asteroid Belt, containing debris left over from the Solar System's formation.

It was after the discovery of Pluto by Clyde Tombaugh on February 18, 1930, that many scientists began to speculate about the existence of similar smaller objects in this region of space, perhaps large in number. By the early 1990s speculation that more objects existed was confirmed, with the discovery of the Kuiper Belt. The Kuiper Belt is vast, substantially more so than the Asteroid Belt—20 times as wide, and perhaps 200 times as massive.

Having left the Kuiper Belt through collision within it or external forces, which could have equally sent them outward into the depths of space, a proportion of Centaurs journey towards the inner part of the Solar System. The inbound journey may see them travel towards the Sun, but more often than not it sees them thrown into highly unstable orbits, through the gravitational pull of one of the outer planets. Occasionally, the pull from one of the planets is so strong that they become captured into orbits around a planet, ending up as moons.

Beyond the Kuiper Belt, as far distant as 144 billion km from the Sun, lie the scattered disk objects (SDOs). These bodies, like Kuiper Belt objects, have very erratic orbits, with only a passing influence from the gravitational pull of Neptune. Most scientists believe that SDOs were probably Kuiper Belt objects, and as much as some of the objects were sent to the inner part of our Solar System through collision, others were dispatched outward. Several SDOs have been cataloged, most notably Eris, discovered on January 5, 2005, which even has a small moon. Sedna is another SDO, discovered two years previous to Eris, on November 14, 2003. Orbiting beyond Kuiper Belt and the SDOs are the "detached objects," which lie completely outside the influence of Neptune, but with their own highly eccentric orbits.

3. Comets

Composition of a Comet

Comets are the true mavericks of the cosmos. That is to not to downgrade or discount asteroids, but somehow, perhaps steeped in myth and legend, these cosmic wanderers hold a special place in the astronomer's eyepiece, for their mystery and magic still provoke much within us; indeed, they touch on our very being (Fig. 3.1).

To start our cometary journey, as with much cosmic debris, the time has to be wound all the way back to the very beginning, some 4600 million years ago. The creation of our universe spawned many objects, but none more prevalent throughout history than the comet. Comets are often referred to as 'dirty snowballs,' the term first coined by American astronomer Fred Whipple (1906–2004) in 1950. Essentially, that's what comets are. Comet are typically broken down into three different parts: the nucleus, the coma, and the tail.

The nucleus of a comet, at the heart of which is a small rocky core, can span anything from 100 m or so upward to quite colossal proportions. A comet is generally composed of rock, dust, ice, and a collection of frozen gases, such as carbon dioxide, carbon monoxide, methane, and ammonia. There is likely to be a crust of some sort binding much of the material together, and, in doing so, this crust allows the original formation components to stay intact for a long period of time. In essence, bar a collision, the original makeup of a comet from perhaps millions of years ago can remain intact for centuries, until warmth from a star, such as our Sun, heats up the comet, allowing the crust to fracture and thus releasing the entombed ice and other material from the crust.

Most interestingly, a comet's nucleus also harbors organic compounds such as methanol, hydrogen cyanide, formaldehyde, ethanol, and ethane, with more complex components also likely, such as long-chain hydrocarbons and amino acids. Manifesting

© Springer International Publishing AG 2017
J. Powell, *Cosmic Debris*, Astronomers' Universe,
DOI 10.1007/978-3-319-51016-3_3

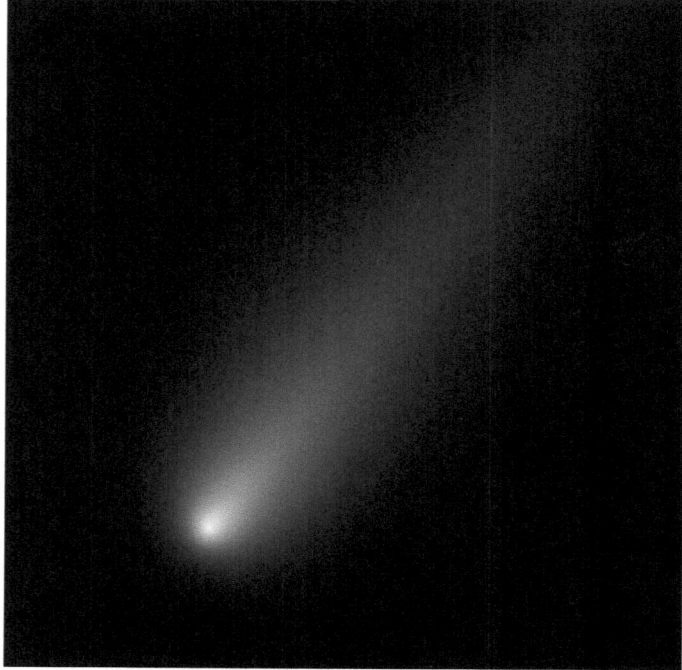

Fig. 3.1 Emerging from the darkness, the awe-inspiring image of a comet (courtesy of NASA)

itself as dark organic material, this, too, is quite an astonishing piece of preservation, made even more interesting by the possibility that it is 'living.'

The second cometary component is the coma, a nebulous envelope that surrounds the nucleus of the comet. The coma is formed as a comet approaches a star such as our own Sun, the crust and ice becoming warm and in time fracturing. The ice eventually turns into gas, forming the coma. Some comets have comas that can reach over 1.5 million km in width. Radiation generated from the Sun pushes the particles away from the coma, forming the third part of the comet's component, the tail.

The tail is illuminated by the Sun, and since comet tails are shaped by sunlight coupled with the solar wind, the ion part of the tail always points away from the Sun, as the ions are more affected than other such scattered dust particles in its composition. As the comet emits streams of dust and gas following its warming, several tails may well form of dust and gas, often quite distinct in nature, with a proportion at quite acute angles.

In some cases, an anti-tail may form, apparently in some cases seeming to cast the impression of a 'spike' in front of the comet. This, though, is merely an optical illusion, with the observer actually witnessing debris that has been left behind by the comet but which appears to be in front of it as Earth passes through the comet's orbital plane. Once a comet has rounded the Sun, it will be seen to continue its orbit tail first, the tail always pointing away from the Sun. Some tails have been known to reach literally millions of kilometers. Depending on its size and approach to the Sun, the coma and tail can make for naked-eye objects, and, in history, such appearances have fueled interesting accounts.

Ancient Appraisal of Comets

The ancient Greeks put forward a number of different theories to explain comets, some believing they were celestial bodies, just like the known planets of the time. Roman philosopher Lucius Annaeous Seneca (4 BC to 65 AD) subscribed to this idea, thinking that the main difference between planets and comets was simply the greater orbits that comets commanded.

One account from this time period gives the explanation for the movement of comets among the stars as more in line with celestial mechanics, such as planetary conjunctions. Another seemingly bizarre but completely plausible explanation at the time placed the answer to any comet's appearance very much closer to home, as some kind of optical phenomenon taking place in Earth's atmosphere not yet fully explained.

Greek philosopher Aristotle (384–322 BC) believed that comets were in fact dry exhalations of Earth that caught fire high in the atmosphere, or similar exhalations produced by the stars and planets. It was this explanation that proved the most favored of all, and was duly adopted until challenged much later by Danish astronomer Tycho Brahe (1546–1601) in 1577.

Brahe argued against the idea by using the parallax method, triangulation (viewing an object from several different perspectives), on a bright comet of the time. Brahe, although failing to measure any parallax, did however determine that the comet was very far away, of a great distance, and at least four times farther than the Moon.

Johannes Kepler, one of Brahe's students, would subsequently devise his three laws of planetary motion, much of which was based on Brahe's extensive observations of Mars. However, despite using the exhaustive work carried out by his teacher, Kepler was still unable to fit comets into the laws of planetary motion modeling, with their eccentric orbits at odds with his notion that comets traveled in straight lines throughout the Solar System.

Aristotle had maintained that Earth was at the center of the universe, with the Moon, the Sun, and other planets in the Solar System all orbiting around it. Given this premise, there seemed to be no room to explain any other phenomena, so a comet had to be Earth-bound, in our own atmosphere. This view also fitted nicely with the model presented by Aristotle, making the appearance of any comet in the sky more earthly in nature, with no outside influence from the universe. As we know, Aristotle's belief was to be commonly adapted and accepted, but it was to prove a double-edged sword.

For many, the wonderment of when a comet now appeared would be taken away by the explanation, a simple dry exhalation! Also this awe was tempered by fear, a fear that the appearance of a comet in the sky might have a meaning, a prophecy in the heavens. Why show this exhalation at particular times and in particular years? What could the appearance of a comet mean? The fear and superstition linked to comets was slowly established—a link that would hound comets for centuries.

Seneca had been one of maverick voices to speak out against Aristotle's theory, claiming in his *Naturales Quaestioes* that the various interpretations of comets, be they atmospheric appearance or otherwise, were old news, claiming that this notion had already been developed by Chaldean astronomers. Chaldean was a term for a so-called settlement within the Babylonian community, set aside for philosophers and whose main intellectual pursuit was astronomy. However, some of these Chaldeans were not approved of within their own group, as they professed to writing horoscopes, something frowned upon as non-scientific. In the eyes of others, Chaldeans collectively were not classed as astronomers at all, just priest-scribes, specializing in astrology and other forms of divination.

Whatever was thought of the Chaldeans, they were considered to be among the first to make systematic observations of the night sky. Their observations and subsequent findings, made from the ancient city of Babylon in modern day Iraq, are cataloged on vast quantities of clay tablets. Dating from the third millennium BC, the Bronze Age, many are very detailed in nature. Seneca had long argued that the Chaldean astronomers had taken a lengthy interest in the observation of unpredictable objects in the sky, with sophisticated ideas about their nature and origin, and that, therefore, Aristotle's view was not new at all!

The Chinese were also fascinated by comets, with their ancient astronomers recording and collating many observations. Some of this early work represents the greatest body of ancient observations on comets ever found, findings dating back to the eleventh century BC.

Another ancient source, known as the Mawangdui Silk Texts, is an almanac in several parts transcribed on silk, dating back to the fourth century. Within its lavish documentation lies incredibly detailed observations of comets made by astronomers, from how these objects appeared in the sky to the observer, their shape, form, and tail, to the supposed orbit and path that it took. However, as prolific and brilliant as this book was, like Aristotle's account of the comets, it also included additional records, where disasters and catastrophes had been linked to the appearance of each comet.

A New Understanding of Comets

With Brahe and Kepler unable to forge a solid explanation for the behavior of comets, it wasn't until English scientist Sir Isaac Newton (1642–1727) used his new law of gravity that a credible solution to the problem was to be put forward. Newton was able to calculate a parabolic orbit for a comet that appeared in 1680. With colleague and friend Sir Edmund Halley (1656–1742), the pair set about defining and explaining the orbits of comets, also accounting for their one-off appearance or subsequent reappearance.

Bearing in mind the comet of 1680, Newton proposed that a parabolic orbit be open, with an eccentricity of 1 (a one-appearance comet never to return), with a circular orbit comprising an eccentricity of 0. Any comets with a less eccentric orbit would be classed as closed ellipses (a regular oval shape), meaning that a comet would return.

Using this method, Halley set about working out the orbits of 24 comets that had been observed, applying the principle to each one. However, with observational records of these comets not being that detailed, leaving a lot of room for speculation, the outcome saw Halley unable to place any of them in separate categories, and thus unable to break down the comets into having one specific orbit or none at all, just passing through.

Halley published his observations relating to the 24 comets in 1705 and, undeterred by the rather vague and perplexing outcome, continued his work. However, during the study of the 24 comets, Halley had interestingly enough noted that the comets of 1531, 1607 and 1682 did have orbits that made them appear at approximately 76-year intervals, pointing to the fact that not only was this probably the same comet but, given the years between its appearance, it was possible to predict the next visit.

Halley therefore predicted that this comet would reappear during 1758, which it did, perihelion occurring on March 13, 1759. Although this was recognized as the first periodic comet, sadly, Halley did not live long enough to see his prediction come true. His observations laid the groundwork for years to come in the field of comets.

It wasn't until the work of Johann Encke (1791–1865) that Halley's periodic comet had a secondary endorsement. Encke, following on from the work of Halley, made a study on a comet that had been discovered by French astronomer Jean-Louis Pons (1761–1831) in 1818. Encke determined that this comet did not seem to follow a parabolic orbit, suggesting that the comet was indeed following the path of a closed ellipse. Furthermore, with an orbit of 3.3 years, it had possibly been observed before on many occasions, but not established as the same body.

Encke recounted observations made by French astronomer Pierre Mechain (1744–1804) in 1786 and German astronomer Caroline Herschel (1750–1848) in 1795, and another much earlier sighting of the comet also made by Pons in 1805. Encke concluded that this comet was the same comet and, with future predictions confirmed, the comet was duly tagged with his name.

Further observations were to unravel a mystery that had been puzzling Encke surrounding the reason why, during subsequent visits, the orbit of the comet appeared to be shrinking, with each return shaving several hours off its previous rendezvous. Astronomers were able to notice such a small shortening of the orbit first because of the frequency of visits made by it and second, and more important, because of Newton's law of gravity, which takes into account other potential influences from planetary bodies.

The measurement of the decaying orbit was one affair, but it didn't answer the question as to what was causing the comet to slowly wind itself towards the Sun. On examination of data on other comets, it was found that Encke's comet and its depreciating orbit were not alone. Other comets, too, including Halley's, were also showing similar traits, with the conclusion that some comets were either arriving too late or too early. Was some as yet undiscovered force acting upon the comets, one that reached not just deep into the Solar System but way beyond? Or was it just the comet itself—loss of mass, perhaps?

German mathematician and astronomer Friedrich Bessel (1784–1846) suggested that the expulsion of material from a comet near to perihelion was acting like the motor of a rocket, each time when rounding the Sun propelling it into a different orbit. Bessel was right, his explanation being confirmed in later years by the improvement in technology and the ability to calculate orbits with more accuracy.

Apart from Halley and Encke, there have been many other sightings of comets that, over time, have turned out to be the same comet, just seen by different people. One such instance is that of the Biela's Comet. First recorded by French amateur astronomers Jacques Lebaix Montaigne (1716–1785) and Charles Messier in 1772, then later sighted again by Jean-Louis Pons in 1805, it was eventually shown by Austrian military officer and astronomer Baron Wilhelm von Biela (1782–1856) in 1826 that all of the sightings that had been made were, in fact, of the same comet.

Biela, having identified that the comet had an orbital period of around 6.6 years, predicted that it would return in 1832, and it duly did so. After a poor apparition in 1839, the comet was recovered in the skies on November 26, 1845, by Italian astronomer Francesco de Vico (1805–1848). At the time of sighting, and due to Biela's faintness, nothing seemed wrong. However, observations made by naval officer Matthew Fontaine Maury (1806–1873) on January 14, 1846, noted that the comet was not alone.

It became apparent that the comet had broken up into two separate components during its journey, both fragments still traveling together along Biela's predicted orbital path. The two portions of the comet (now named Comet A and Comet B) traveled successfully around the Sun and journeyed away as they had arrived.

Biela's Comet was rediscovered on its return in 1852 by Italian astronomer Angelo Secchi (1818–1878)—first Comet A then

Fig. 3.2 Comet 73P/Schwassman-Wachmann 3 in a state of disintegration. Photograph captured by NASA's Spitzer Space Telescope (courtesy of NASA)

later Comet B. Following this visit, the double comet was never seen again.

Having been named Biela's Comet, its breakup and subsequent disintegration left a serious question mark over what had happened to it and why it had happened. To mark the comet's demise, a glorious shower of meteors, boasting an hourly rate of 3000, lit up the skies in November 1872, more or less where the comet was supposed to have reappeared. The Bielid shower is more commonly known as the Andromedids, as they radiate from the constellation of Andromeda (Fig. 3.2).

Early Accounts of Halley's Comet

Halley's Comet is the most famous of all periodic comets, with its visitations to Earth making it one of the most documented of all heavenly bodies (Fig. 3.3).

Cometary references by ancient astronomers are few. This is not to say that records were not readily kept on comets, just that the passage of time has deprived us of such detailed accounts, with perhaps many sightings still yet uncovered. A brief Chinese account dating from 240 BC remains the earliest identifiable record of Halley's Comet, before the Babylonian tablets of 164 BC and 87 BC.

A record of the 240 BC return of the comet indicates an appearance in the eastern aspect of the sky between the dates May 24 and June 23. After conjunction, a further 16 days of observations on the comet were collected during its western apparition. It is quite likely that Babylonian tablets from this year exist, but are yet to be uncovered. The earliest credible sighting must therefore be accredited to the Chinese. Perihelion for the 240 BC return occurred on May 25, with the comet's closest approach to Earth measured at a distance of under 67 million km. The next return saw Halley's Comet come even closer, at under 16 million km.

There are other references to a comet made much earlier in history. A comet was recorded in ancient Greece in 467 BC, more than likely being Halley's Comet. Chinese chroniclers also mention a comet at this time. Despite these references, which seem to be irrefutable, the first certain appearance is still that of 240 BC.

Fig. 3.3 Reproduction of original plates of Comet Halley, May 25, 1910 (courtesy of ESA)

The Chinese logged accurate and extensive data on the return of the Halley's Comet from 240 BC onwards, but Western records lag significantly behind, until, that is, the return of the comet in 1456, and the subsequent work on this apparition by Italian astronomer Paolo Toscanelli (1397–1482). Toscanelli's diligence in observing the comet over a month and calculating its positional measurements to within a degree marked a significant shift in accuracy, away from the traditional style of observation used by Oriental astronomers to a more precise regime.

That said, the early work by the Chinese, Japanese and Koreans with regard to Halley's Comet remain utterly irreplaceable. These earlier calculations afford us valuable insight into the orbital data for the comet, with documentation noting several approaches to Earth. One such occasion brought the comet around 5 million km away from Earth, and it is these engagements that have undoubtedly altered the motion of the comet over time.

To reinforce the importance of early Chinese observational work, in 1846, English astronomer John Russell Hind (1823–1895) became one of the first to extensively use early observational data in an attempt to make orbital predictions more accurate. Subsequent astronomers followed this method, with British astronomer Phillip Herbert Cowell (1870–1949) and French-born astronomer Andrew Claude de la Cherois Crommelin (1865–1939) using such reckoning for the comet's return of 1910. Later, Tao Kiang (1929–2009) of the Dunsink Observatory in Dublin and Donald Yeomans of the Jet Propulsion Laboratory in America were to also use the records.

The 198 BC return saw Halley's Comet first sighted in the constellation of Taurus to the east, in the area of the Pleiades, and later sighted to the west in Sagittarius between October 21 and November 19. There is no reference on either Babylonian tablet as to the dates of the comet's first sighting to the east, but the second sighting dates are mentioned, with a general idea of when it was probably seen.

The 87 BC return of Halley's Comet was at a distance more in line with that of 240 BC return, with a closest approach measured at less than 67 million km. Chinese accounts at the time recollect the sight of a comet in the eastern aspect of the sky. The Han-shu records reference the comet thus: "In autumn during the seventh month (August 10 to September 8) there was a 'bushy star' in the east." The Babylonian account, despite coming from only a fragment of a tablet, talks of the comet's tail, with its position noted sometime between July 14 and August 11, perihelion occurring on August 6.

The 12 BC apparition of the comet was chiefly documented by the Chinese, Babylonian records seemingly having not survived the passage of time, although it is fair to say that it is likely that an account was made. The motion of the comet during this return is accurately accounted for by the Chinese, as it moved from Gemini to Scorpius. The comet's first appearance was documented on August 26; its last charted appearance before fading out of sight was noted on October 20. These exact dates give a total duration of visibility for Halley's Comet during this visit at 56 days. The comet's closest approach to Earth occurred within 24 million km, with perihelion on October 10.

The return was considered among a number of portents preceding the death of Roman statesman Marcus Vipsanius Agrippa (63–12 BC). An account at the time reads: "The star called comet hung for several days over the city and was finally dissolved in flashes resembling torches."

Halley's Comet next returned in AD 66, first sighted on January 31, with the last documented sighting on April 10, having passed through several zodiacal signs. Perihelion occurred on January 26, with the comet's closest distance from Earth measured at within 37 million km.

During this apparition, Romano-Jewish scholar and historian Titus Flavius Josephus (AD 37 to AD 100), who lived through the siege and fall of Jerusalem, makes reference to the comet's apparition. He mentions several portents leading up to these events, one of his descriptions noting a phenomenon in the sky, possibly referring to Halley's Comet. Although only speculation exists about what he was actually referring to, it seems a distinct likelihood that it was Halley's.

During the AD 141 return, references made to it by Chinese records on the color of the comet detail it as "pale blue." The Chinese also made a very accurate measurement of the comet's nucleus. First sighted on March 27 to the east, last visible on April 22 to the west, perihelion occurred on March 22, with the comet's closest distance to Earth measured at under 25 million km.

For the AD 218 return of Halley's Comet, we rely mainly on the Chinese for records of any substance, with the first sighting noted on April 14. Visible in the western part of the sky, the comet's motion is logged as being sighted between the constellations of Auriga and Virgo. Perihelion occurred on May 17, with the closest distance from Earth measured at within 64 million km. There is reference to the comet being sighted from Europe during this return in the contemporary work of Lucius Cassius Dio, a Roman statesman and historian. Dio, who was to publish 80 volumes of his *Roman History* during his lifetime, lists the comet as a portent before the death of Emperor Caesar Marcus Opellius Severus Macrinus Augustus (Emperor Macrinus, AD 165 to AD 218).

A fairly unremarkable return was noted by the Chinese for AD 295, strangely sketchy in any detail, with only the month of discovery (May) documenting the comet's appearance. First seen in the constellation of Andromeda, last seen in Leo, it is not known whether the lack of information on the return is because of a particularly poor apparition of the comet or whether records of any substance simply haven't survived. Equally, and not impossibly, sustained poor weather conditions may have been to blame.

A very close approach of Halley's Comet occurred in AD 374, with the comet literally skimming Earth at just over 12 million km at its nearest encounter. At this distance, the comet must have presented itself with one of its most splendid appearances in our skies, with Chinese records referencing a "bushy star" sighted in Aquarius on March 6, then a "broom star" sighted in Libra on April 2. Another comet is also documented at this time, which probably goes some way to confusing records of the period. However, this other comet seemed to have different motions, and, as there are references to this second comet appearing in a different parts of the sky, we can rule it out as being associated with Halley's in any way.

The AD 451 return of Halley's Comet is well documented. This apparition also coincided with the time of the Battle of Chalons, in which Roman general Flavius Aetius (391–454) engaged his 50,000-plus troops against Attila the Hun (406–453) and a similarly numbered compliment of men. Attila was defeated at the battle. References are made to a comet being seen around June 10.

This return also saw, according to documents, the comet in the skies for more than two months. However, it was not a particularly near encounter with Earth following the comet's previous return, some 74 million km at closest approach during this apparition, with perihelion on June 28.

From its "pale blue" appearance of AD 141 to a "pure white" offering during the return in AD 530, the Byzantine chroniclers of the time describe what they saw in September of that year as a "huge and terrible star" whose rays (tail) extended towards the zenith. Visible for a total of 20 days, the chroniclers referred to the comet as Lampadias, as it resembled a burning torch in the sky.

Halley's Comet next appeared in AD 607, and despite its distance from Earth measuring just over 12 million km, in line with the approach of AD 374, accounts of the return from the Chinese and from observers in Europe lapse into another vague phase of documentation, with the comet on view for what appears to be two months. There are also references to other comets being sighted at this time.

Although the AD 684 return seems virtually unreported across Europe, it was documented for the first time by the Japanese, with reference made also to a different comet sighted at this time. There is a reference made to this other comet in the *Life of Pope Benedict II* (635–685). The account refers to the comet as being "an absolutely shadowy star, resembling the Moon when covered with clouds." Halley's Comet was first depicted in the *Nuremburg Chronicle* on this return, although there is speculation with regard to a link between this sighting and the date of the *Chronicle*. There is a possibility that there are other, earlier, recorded sightings in the *Chronicle*, but the AD 684 drawings do seem to tie in with this apparition.

First sighted by the Chinese, and indeed the Japanese, in September 684, the comet made for a spectacle the western sky, with an impressive tail of some 10°.

Referred to as a "glittering beam" by Byzantine chroniclers, the AD 760 return saw the comet on view for 50 days—first sighted in Aries, then much later in Virgo. With the sighting of another comet also noted during this time, this particular appearance of Halley's Comet is regarded as having serious astrological consequences, especially since later in the year, on August 15, a solar eclipse occurred. The sight of two comets at once caused a great deal of disquiet, perhaps because of the impressive spectacle that it created, with both Halley's Comet and the other comet viewed in separate parts of the sky.

From under 61 million km at closest approach to Earth in AD 760, the return of AD 837 was to see Halley's make its closest known approach in history, barely 10 times the Moon's distance, at just over 5 million km.

Earth's gravitational field had a distinct influence on the comet on that visit, and was perhaps responsible for the comet's trajectory being altered, so much so that it might never make such a close approach to Earth again. Although other outside influences may still reshape future returns of Halley's Comet, the encounter of AD 837 must still have been a magnificent sight, with the comet speeding across the sky at around 2° an hour, sporting a tail spanning some 60°, with a distinct second tail also visible.

With observations made throughout Europe and the East, this encounter is also a significant reminder that, even if periodic comets follow a reasonably stable orbit, gravitational forces from other bodies can have significant influence in altering orbital paths. If Halley's Comet on this return had arrived a week earlier, it would have passed between Earth and the Moon, and, at this distance, it could have caused a radical shift in the orbit of the comet, although only conjecture can surmise whether that effect would have been a positive or negative influence on future Earth encounters.

The return of Halley's Comet is not well referenced in AD 912, the Chinese completely missing it for some reason, with the only documented evidence of the time from the Japanese. However, there is a reference to not one but three comets by Baghdad chronicler Ibn al-Jawzi (AD 1116 to AD 1201), an Islamic scholar and prolific writer, author of over 700 books.

Despite the Chinese rediscovering the comet in AD 989, it is an Arab chronicle record that notes a "planet with a tail," sighted and tracked over Cairo for 22 days.

The Chinese Contribution

The records of the Chinese continue after AD 989, with further documented sightings of not just Halley's Comet but other bodies, as their significant contribution to the keeping of historical records continued.

There was still a lot to be gleaned from their early observational work, but it wasn't until 1843 that former railway worker and French Orientalist Édouard Biot (1803–1850) made an extensive translation of these ancient Chinese astronomical records. English astronomer John Russell Hind was one of many to make use of the work of Biot, attempting to correlate observations that had been made by the Chinese every 75 years, roughly the orbital period of Halley's Comet. In doing so, Hind strived to see just how far back he could go with the Chinese data, and subsequently it is Hind that we have to thank for linking the image of Halley's Comet to the one on the Bayeux Tapestry.

Despite some anomalies in the research conducted by Hind, essentially he did revolutionize the whole history of the comet. His work was progressed in later years by British astronomer and specialist in the field of mathematics Phillip Herbert Cowell (who also notably discovered asteroid 4358 Lynn) and Andrew Claude de la Cherois Crommelin, also a British astronomer and president of the Royal Astronomical Society from 1929 to 1931. Crommelin, considered at the time to be one of the world's leading authorities on comets, worked closely with Cowell and made predictions based on the work of the Chinese and other available records, correctly forecasting the return of Halley's Comet in 1910 to an accuracy of just over three days.

Crommelin's other work also concluded that Comet Forbes III, Comet Coggia-Winnecke 1873 VII, and Comet Pons 1818 II were in fact the same periodic comet, sighted at different intervals during its appearance to Earth-bound observers. The comet was subsequently renamed 'Comet Pons-Coggia-Winnecke-Forbes'!

Latter-Day Accounts of Halley's Comet

Over the centuries to follow, there are notable landmarks in the extraordinary catalog of Halley's Comet, with possibly the most referenced being that of AD 1066, for here, the Bayeux tapestry was to capture Halley's Comet in all its glory, depicting an ailing Harold on the throne pointing to his eye, as his surrounding courtiers gesture towards the comet in absolute awe and terror. The caption to accompany the scene read *"isti mirant stella,"* "they wonder at the star." Although William of Normandy had been preparing his invasion of Britain, he is alleged to have remarked, "A comet like this is only seen when a kingdom wants a king."

According to the Anglo-Saxon chronicle, "Then there was seen all over England a sign such as no one had even seen. Some said that the star was the comet-star, which others denominate the long-hair'd star. It appeared first on the eve called "Litania major," that is, on the eighth before the calends off May; and so shone all the week."

From around the globe an extensive catalog of sightings exist for the for AD 1066 return. The cathedral archives at Viterbo, in Italy, record the comet as having a tail streaming like smoke up to nearly half of the sky. The comet is also noted again by Baghdad chronicler Ibn al-Jawzi, its reappearance after conjunction described in his writings: "It reappeared on Tuesday evening at sunset with its light folded around it like the Moon. People were terrified and distressed." The Koreans, who were making great advances in their recording of astronomical events, referred to the comet's appearance as being as large as the Moon.

The Chinese documented the first sighting of Halley's Comet in the AD 1066 return on April 3, with a total duration of observations spanning 67 days. Their description of a "broom-like" vapor is just one of many accounts from the East, with local chronicles across Europe and Russia widely reporting the motion of the comet over the months.

Since the Bayeux tapestry, Halley's Comet has been further depicted many times. In AD 1145, a drawing of the comet appears in the *Eadwine Psalter*, an elaborate and beautifully illustrated manuscript written and attributed to Eadwine, a monk and scribe at Canterbury Cathedral during the twelfth century. In the book, the comet is depicted at the bottom of the page where Psalm 5 begins, and it is believed that Eadwine sketched the comet, because it appeared in the sky as he was completing the page that he was working on.

Preserved at Trinity College, Cambridge, his work has an Old English comment besides the depiction of the comet that, when translated, states: "Concerning the star called comet. A suchlike ray has the star known as Comet, and in English it is called 'the hairy star.' It appears seldom, after [periods of] many winters, and then for an omen."

A further notable depiction is that of the comet's return of AD 1301. This apparition was captured by the Florentine painter Giotto di Bondone (1266–1337), in his painting "The Adoration of the Magi," in the Scrovegni Chapel at Padau, Italy. Painted in 1303, it shows the Star of Bethlehem as the comet.

Giotto Space Probe

During the 1986 return of the comet, the painter Giotto was commemorated by the European Space Agency (ESA), which sent a probe to encounter not just Halley's Comet but also another comet, Grigg-Skjellerup (Fig. 3.4).

Giotto reached Halley's Comet on March 14, 1986, going on to observe Grigg-Skjellerup on July 10, 1992.

This comet flyby was an historic mission, the first to make close-up observations of a comet, firstly closing in at a distance of just 595 km to Halley's Comet, then even closer, to within 200 km of Grigg-Skjellerup.

Fig. 3.4 Giotto spacecraft (courtesy of ESA)

Giotto was part of a fleet of probes sent to encounter the 1985/86 return, with Soviet Vega probes 1 and 2 also contributing valuable data. Referred to as the "Halley Armada," Vega 1 started the combined mission with a rendezvous and subsequent transmitting of images on March 4 1986, the first ever of a comet's nucleus. Vega 2 encountered the comet on March 9 1986, at a distance of 840 km.

Also parts of the Armada were two Japanese probes, Suisei (Japanese for comet) and Sakigake (translated as pioneer or pathfinder). As the name suggests, Sakigake was the forerunner for Suisei, sent ahead as a practice mission.

Giotto and the two Vegas were a triumph of the time, and despite being battered by cometary dust particles and debris, images produced likened the comet's nucleus to a dark, dust-covered, peanut-shaped body, 14 km long, varying in width from 6 to 9 km. The formation of the comet was dated to 4.5 billion years ago.

Much of the nucleus was inactive, with only 10% showing some activity. Three outgassing jets were photographed on the sunlit side of the nucleus. Giotto was able to measure the volume of material being ejected from Halley as being 80% water, 10% carbon monoxide, and 2.5% a mix of methane and ammonia. Trace amounts of hydrocarbons, iron, and sodium were also detected. Any material left by Halley during this encounter subsequently fueled future occurrences of the two showers associated with the comet, the Orionids and the Eta Aquarids.

Halley—The Prediction He Never Saw

Having observed the comet for himself from August 26 to September 10, 1682, Sir Edmund Halley, the second Astronomer Royal, had noted how similar this comet's path was to those observed in 1531 and 1607. Halley concluded they were one and the same, predicting the return of the comet "about the end of the year 1758 or the beginning of 1759." The comet duly reappeared in the skies as predicted, first sighted on Christmas night 1758, coming to perihelion on March 13, 1759.

Halley, having made the prediction, knew he would not live long enough to see the comet reappear in our skies, and 16 years before its return he passed away, on January 14, 1742.

Of the comet, Mark Twain said, "I was born with Halley's Comet and expect to die on its return." Twain passed away on April 25, 1910, two days after the comet's perihelion.

Following the apparition of 1835 (perihelion on November 16, 1835), Halley's Comet was next to return in 1910, being first sighted a year previously on September 12, 1909, by Max Wolf in Germany.

At a distance of 40 million km from the Sun, the comet sped towards its perihelion on April 19, 1910, passing between Earth and the Sun on May 18–19. Traveling at 5 million km a day, Halley's Comet put on a glorious display on the morning of May 19 when, after the Moon had set, the tail was seen to stretch for 120°, a full two-thirds of the way across the sky. On this particular day, it is thought that Earth actually passed through the comet's tail. By June 5, 1910, the comet had lost its plasma tail, leaving the original tail far behind Halley's track as it embarked on the outward leg of its orbit.

The 1986 return was a less than favorable visit by our cosmic companion, putting in a rather disappointing appearance, but, even so, Halley still conjured up much enthusiasm across the globe. First sighted with the 200-inch Hale Telescope at Palomar Observatory on October 16, 1982, the comet sped to perihelion on February 9, 1986, passing Earth at a distance of 62 million km on April 10.

Halley's Comet will next enter the inner Solar System in the year 2061. This, too, is not expected to be an overly favorable visit observation-wise.

Rosetta Space Probe

Giotto's encounter with Halley's Comet was an astonishing achievement, with subsequent missions to comets always on the agenda as future targets. In more recent times, the ESA's Rosetta Mission was an equal triumph. Launched on March 4, 2004, Rosetta, along with its lander module Philae, embarked on a

Fig. 3.5 An artist's impression of Rosetta descending to the surface of Comet 67P Churyumov-Gerasimenko, September 30th 2016 (courtesy of ESA/NASA)

journey and eventual study of a comet in a mission that was to last 12 years, 6 months, and 28 days (Fig. 3.5).

Launched on the back of an Ariane 5 rocket from French Guiana, Rosetta reached its target, Comet 67/P Churyumov-Gerasimenko, following earlier scheduled encounters with Mars and two asteroids, 21 Lutetia and 2867 Šteins. Rosetta became the first spacecraft to orbit a comet, spending two years collecting data about its composition and structure, also managing to successfully deposit the Philae lander for the first controlled landing on a comet's nucleus.

Comet 67/P Churyumov-Gerasimenko is a Jupiter-family comet, thought to have entered the Solar System after leaving the Kuiper Belt. However, this was not the first choice comet as a target for Rosetta, with an earlier launch of the spacecraft planned for January 12, 2003, canceled due to complications with the Ariane rocket. The scheduled encounter with Comet 46P/Wirtanen in 2011 was subsequently abandoned until a new launch window could be designated, and a new comet to rendezvous with found (Fig. 3.6).

With over 100,000 images sent back to Earth, the task of sifting through the data is an enormous one, but also a potentially revealing one. Already scientists are musing over the discovery of

Fig. 3.6 Picture of the surface of comet 67P Churyumov-Gerasimenko captured by the navigation camera on the European Space Agency's Rosetta Spacecraft in October 2014 (courtesy of ESA/NASA).

what some have referred to as "goose bumps" or "dinosaur eggs," 3-m lumps of material deposits on the comet's surface, which could be evidence of building blocks from which the comet was eventually formed.

Although Philae also made the discovery of water on the comet (a find that was expected), the water was unlike that found on Earth, which prompted the science community to rethink the theory that water came to Earth from comets. With comets not totally ruled out and more observations required the theory of how water came to Earth rests more with asteroids than comets, unless different comets hold different types of water.

With the spacecraft power failing, Rosetta was finally crashed at walking pace into the very comet it had been studying, ending another significant and successful chapter in the quest for an understanding of comets.

Great Comets of Our Time

The Great Comets of 1811 and 1843

The Great Comet of 1811, until the appearance of Comet Hale-Bopp in 1997, held the record for the longest amount of time that a comet had been visible to the naked eye, a staggering 260 days. With an orbital period estimated at 3100 years, this comet attained an apparent magnitude of 0 at its brightest, and was thought to have had a significantly large coma, perhaps reaching over 1.5 million km across, 50% larger than the Sun, although this estimate has been noticeably downsized over the years. Sometimes referred to as Napoleon's Comet, the Great Comet of 1811 is popularly thought to have portended his invasion of Russia, and the War of 1812.

References are also made to the comet's appearance as a daylight object during the New Madrid earthquakes in December 1811. Also, at the midpoint of *War and Peace*, Tolstoy describes the character of Pierre observing this "enormous and brilliant comet [...] which was said to portend all kinds of woes and the end of the world."

After its discovery on February 5, the Great Comet of 1843, also known as the Great March Comet, brightened rapidly to become a naked-eye object during the early part of the year, with its perihelion on February 27. A member of the Kreutz sungrazers family of comets (see Ikeya-Seki), its closest approach to Earth was on March 6, and its greatest brilliance was achieved the following day. Skimming the Sun during its perihelion at a distance of under 830,000 km, the Great Comet at that time had passed closer to the Sun than any known object.

Comet Daniel

Discovered by Zaccheus Daniel on the morning of June 9, 1907, Comet Daniel was to become the most photographed comet to date. The comet's arrival and subsequent photographing firstly showed how far the photographic world had advanced, but more importantly it was an indicator that the sky was now open to more ways of being observed than ever before, and generally by more people.

Using a 6-inch telescope at the observatory in Princeton, New Jersey, the comet was picked up in the eastern sky near to Saturn. From mid-July, it had brightened sufficiently to be seen with the naked-eye, staying on view to the unaided observer until mid-September. Observed low, just above the horizon before dawn, Comet Daniel swung into the western sky, sporting a tail that had steadily developed over the previous months, stretching 15° at its greatest length.

It is thought that Comet Daniel was subject to much influence from the planets in our Solar System, particularly those of the outer planets: Jupiter, Saturn, and Uranus.

The Great January Comet

Along with Halley's apparition, there was another spectacular comet in 1910, that of the Great January Comet, also known as the Daylight Comet. First sighted in the southern hemisphere on January 12, numerous observers were to claim that they were the first to sight the comet, and should duly have their name tagged to it. However, no single person has ever been attributed with the find, although its sighting in the pre-dawn sky by diamond miners in the Transvaal is widely quoted as being the first visual contact.

That said, this, too, is open to question, as by the time it was sighted on the morning of January 12 the comet had noticeably brightened (albeit rather suddenly), already reaching an apparent magnitude of −1, making for a prominent object in the sky, and thus availing itself to many who cared to cast their gaze upwards.

Following a tip-off from the editor of a Johannesburg newspaper on January 15, Scottish astronomer Robert T. A. Innes (1861–1933) was able to turn his attention at the Transvaal Observatory to the comet two days later on January 17, giving this new visitor to our region of space the careful scientific analysis it merited as it reached perihelion.

Observers across the southern hemisphere were treated to a glorious spectacle, as the comet blazed a trail. The midday sky revealed the comet in all its magnificence, despite being just a few degrees away from the Sun. To the naked eye, the comet's brilliance was likened to that of Venus at its best apparition, according to some, at least five times the brightness. In telescopes, the comet revealed a fan-shaped head with a short curving tail, with the bizarre sight of a blue daytime sky for a background.

Following the days after perihelion, the comet, having made its way into the eager gaze of northern hemisphere observers, now became a glorious evening object, sporting a vast, extensive and broad tail in the fading twilight, a tail that stretched 50° by late January and into early February. Around January 20, telescopes revealed that the comet's tail had split into two distinct branches, extending well beyond the edges of the eyepiece field. By the end of the first week of February 1910, the comet's glory days had faded, spiraling down to 4th magnitude, but, despite this, observers still reported the presence of a 20-degree ghost-like tail still visible to the unaided eye.

With the return of Halley's Comet predicted for the spring of 1910, the Great January Comet did lead to some confusion at the time. The latter was decidedly brighter and easier to observe than the former, but with such a spectacle comets had well and truly entered the public domain in a big way.

Even in present-day observations, unless a passing phenomenon in the night sky is readily visible for all to see, it does not fire the imagination of the masses, probably because images and talks given in the media simply can't replace the sight of seeing something for oneself, which is why Halley's naked-eye appearance in 1986 did so much for astronomy globally, despite it being such a generally poor showing.

In 1910, astronomers added to the confusion between Halley's Comet and the Great January Comet by publicly announcing the discovery of cyanide in Halley's tail, a discovery made by the use of spectroscopy. Spectroscopy, which was only understood within the confines of those who studied it, had inadvertently opened a can of worms. With Earth due to pass through Halley's tail on May 18, the media, growing in stature and reach during this time, seized on the revelation. However, despite the furor, all was to pass without incident, but it certainly fueled the prophetic concerns echoed from history.

Comet Beljawsky and Comet Brooks

Two notable comets made an appearance in 1911: Comet Beljawsky and Comet Brooks. Discovered on September 29, 1911, in the dawn sky by Russian astronomer Sergei Ivanovich Beljawsky (1883–1953) working at the Simeiz Observatory on the Crimean peninsula, this comet attained naked-eye status and was widely observed during its apparition, with early October seeing the comet visible both in the morning and evening sky.

The evening sky was to provide another spectacle alongside Comet Beljawsky, as Comet Brooks made for the rare double of having two comets on view at once. Discovered in July 1911 by British-born American astronomer William Robert Brooks (1844–1921), this comet, too, achieved naked-eye status, with the evening sky of October 11 revealing both comets, around 20° apart in the twilight.

From mid-October, Comet Beljawsky sat low on the western horizon, displaying a golden-yellow head with a tail stretching 15°. Comet Beljawsky, whose discoverer also found 26 minor planets in his lifetime, reached +1 in magnitude, with Comet Brooks reaching +2. Together over the horizon they must have made for quite a sight, the contrast in tails most notable, with Comet Brooks sporting a blue, narrow, straight tail of 30°, denoting the presence of carbon monoxide ion emissions.

Comet Mellish

Early April 1917 saw the discovery of the next notable comet, but this was a comet with a difference. Discovered by John Edward Mellish (1886–1970), this comet remains of interest not because of any particularly startling appearance made in our skies, but because of the man who made its discovery.

This is an early case of people outside of the scientific community making a find, and bearing in mind the advances in optical aids since that have allowed many others to make similar such discoveries, a landmark feat. Mellish, who lived in Madison, Wisconsin, was an amateur astronomer and telescope maker and during his time he was to find another five comets, bringing his total haul to six.

For his work, Mellish received astronomical medals from both America and Mexico, all from discoveries made at the eyepiece of his home-built array of telescopes—a fine achievement. Mellish was also the second person after Edward Emerson Barnard (1857–1923), more commonly known as E.E. Barnard, to announce that he had discovered craters on Mars. Despite some controversy surrounding this claim in the astronomical community, a crater on Mars was duly named in his honor.

Comet Mellish was chiefly a southern hemisphere object, with a tail that was estimated to span 20°. Calculations propose a return of this periodic comet in 2061, the same time as Halley's next scheduled return. The 2061 return of Halley's is not expected to be one of the best, so perhaps the man and his comet (Mellish) will have a second chance to shine.

A link has also been established between Comet Mellish and two meteor showers, the December Monocerotids, and the November Orionids. Studies seem to suggest the parent comet for the showers is that of Mellish, with perhaps another shower, the December Canis Minorids, also linked with the comet. The data for the conclusion with the connection between the Monocerotids and the Orionids stems from work carried out by the SonotaCo Network in Japan, whose software in this field of study looks at the distribution and orbital evolution of cometary debris.

Comet Skjellerup-Maristany

Comet Skjellerup-Maristany was discovered independently by two amateur astronomers, John Francis Skjellerup (1875–1952), in Australia on November 28, 1927, and Edmundo Maristany (1855–1941), in Argentina on December 6, 1927.

Sighted in the constellation of Norma and rapidly moving northwards towards the Sun, this long-period comet (36,508 years) was visible in the daytime as an intensely bright object, brilliant yellow in color, caused by the emission of sodium atoms. The comet was visible for 32 days before being lost in the bright evening twilight. During its daytime apparition, if an observer were to block out the Sun, the comet appeared within 2° of the Sun and was still clearly visible.

On the morning of January 15, 1941, while observing a variable star, Reginald Purdon de Kock (1902–1980), based in Paarl, South Africa, discovered a comet with a +6 magnitude southwest of Antares in the constellation of Scorpius, making its way in a southeasterly direction. The findings were reported by De Kock to the Royal Observatory at the Cape of Good Hope. Upon receiving the news, the observatory duly turned its gaze on the comet, which had already developed a tail of around half a degree. The Royal Observatory subsequently confirmed De Kock's discovery, although news of the sighting did not make it outside of Europe because of World War II.

The comet was also sighted in Australia by two other independent amateur astronomers, one of them being John Francis Skjellerup, the name Barnes-Skjellerup now having been unofficially attached to it. Just to confuse the issue over who saw the comet first, John Stefanos Paraskevopoulos (1889–1951), an astronomer in Bloemfontein, had reported the discovery to Harvard College, Cambridge, Massachusetts, with the name Comet Paraskevopoulos having been attached to it. With so many sightings now filed, the comet finally landed the official tag of De Kock-Paraskevopoulos, which must have pleased some but not others.

Reaching a maximum brightness of +2, this once naked-eye comet remained on view until the middle part of September 1941.

But by then it had become a very faint object, +17 magnitude, with the last observation of the comet on September 17 made by the Lick Observatory at the University of California.

The Great Southern Comet

During December 1947, the southern hemisphere evening sky suddenly revealed a comet, brightening rapidly before being engulfed in the twilight. Duly named the Great Southern Comet, this visitor blazed the sky with a distinct orange head and a tail that stretched 30° across the sky.

First sighted on December 7, the Great Southern Comet faded as quickly as it had appeared, and was lost to the naked eye by December 25. Moving eastward across the constellation of Sagittarius, so numerous were the original sightings that no person was credited with the discovery. By January 15, 1948, the comet was more than 209 million km from Earth, still visible in large telescopes as it passed through the constellations of Capricorn then Aquarius.

Originally designated 1947n, as the fourteenth comet dis-covered in 1947 (alphabetically ordered 'a' to 'z' during the year of discovery), 1947 was a record year for the discovery of comets. Apart from the numerous land-based sightings from across the southern hemisphere, one particular observation stands out, one made by the crew of a ship in the Pacific Ocean, clearly visible with the naked eye about 8° above the horizon. This must have made for a fabulous sight with the combined elements of the sea, a crisp clear sky and a bright comet.

It was a great comet, no doubt, but a rather fleeting one, with its magnificence and tail in decline days after its startling appearance. Northern hemisphere observers were to miss out completely.

Several comets have been sighted during solar eclipses, one dating back to AD 47, recorded by the Chinese and Byzantine astronomers, and also the sighting of a comet during an eclipse on April 30. However, of all the eclipse-related comets, the Eclipse Comet of 1948 holds the most prominence.

Formally known as C/1948 V1, this comet was discovered as a brilliant object around 2° southwest of the Sun during a total eclipse on November 1, 1948. Although, as we've established, it wasn't the first to be seen during an eclipse, and with a magnitude of −2, the tag of Eclipse Comet stuck.

Following the eclipse, the comet was next sighted on November 4, as a 0 magnitude object in the morning twilight. Northern hemisphere observers were to mostly miss out again, as the comet, sporting a 20-degree tail by now, moved southeastwards through the constellations of first Hydra then Puppis, its tail reaching an estimated peak length during November and December of 30°.

By late November and only being seen very low down, at best, in northern hemisphere skies, the comet faded and, as its brightness waned, its tail shortened. By the start of December the comet had faded to +3.5 magnitude, declining to +5 by the end of the first week and becoming lost over the days to follow. Interestingly enough, the comet's trajectory pointed towards this visitor probably returning to its origin, the Oort Cloud.

Comet MrKos

Two comets of note were discovered in 1957: Comet MrKos in July and Comet Arend-Roland in November.

Discovered by Czechoslovakian astronomer Antonin MrKos (1918–1996) in the dawn twilight of July 29, 1957, Comet MrKos was to attain naked-eye status, with its initial apparition during late July and early August allowing it to be seen both in the morning and evening sky. First sighted near to the bright star Pollux in the constellation of Gemini, the comet displayed two tails, the brighter one strikingly curved in nature and stretching 15° long.

Apart being associated with many comet finds, MrKos was also to discover 274 asteroids during his illustrious lifetime, his last named 217628 Lugh, in 1990, sighted from the Klet Observatory in the Czech Republic. This particular asteroid, from the Apollo family, is named after the Celtic god of sun and light. It has an orbit of around 4 years and 40 days and is classed as a potentially hazardous asteroid (PHO).

Comet Arend-Roland

In November 1957, Belgian astronomers Sylvain Arend (1902–1992) and Georges Roland (1922–1991) discovered a new comet while looking at photographic plates. The eighth comet to be discovered that year, Comet Arend-Roland, like Comet MrKos, also had two tails, but, bizarrely enough, one of the tails pointed towards the Sun while the other pointed away from it.

This anomalous tail, called an anti-tail, appeared for a few nights in April 1957, apparently changing direction over the hours, with a distinct sharp spike aimed directly at the Sun. After in-depth study, it was determined that what observers were seeing was actually an optical illusion. It was the effect of perspective, whereby the observer was viewing not a spike in front of the comet but actually an edgewise fan of debris strewn out behind the comet along its orbit. As Earth passes through the cometary plane (bearing in mind Comet Arend-Roland's closest approach to the Sun occurred on April 8, 1957), the disc is seen side-on, appearing as a spike ahead of the comet. This anti-tail, which other comets such as PanSTARRS and Lovejoy have also shown, is therefore only visible for a brief period of time, when Earth passes through the comet's plane, typically several days at most.

Comet Seki-Lines

Proving that it's not all flipcharts, chalkboards and computers, the discovery of Comet Seki-Lines in 1962 set a wonderful precedent for the diversity of discoveries by individuals from different backgrounds. First spotted by amateur astronomer Richard D. Lines (1916–1992) on the evening of February 3, 1962, while out observing the sky with his telescope in the Arizona desert, this was the counterbalance to professional Japanese astronomer Tsutomu

Seki, who during his lengthy career not only found 6 comets but also 225 asteroids. Comet-Seki-Lines was Seki's second comet, but the glory would have to be shared with his amateur colleague in America.

Lines, who had made a rule about driving away from the Arizona smog at least once a month to experience fresh air and clear skies, had been observing an object in the constellation of Puppis. After consultation with his star charts, Lines noticed that something was amiss. Having been joined by his wife, Helen, and friends, and after further debate, Mrs. Lines declared that the unaccounted-for object in Puppis looked like a comet!

After many years of searching the skies, this avid amateur astronomer and comet hunter had, unknowingly at the time, struck gold, much to the delight of the assembled party. Using just his 15-cm portable telescope, he had made the discovery of a lifetime. With much haste, Lines quickly drove back to Phoenix and to the Lowell Observatory, where astronomer Robert Burnham Jr. (1931–1993) was at work. Burnham confirmed to Lines that he had discovered a comet.

Meanwhile, shortly before midnight in Japan, Tsutomu Seki was also observing the same constellation that Lines had been scouring, taking in a misty patch in the southern Milky Way. Initially having also sighted the object, Seki was torn as to whether or not it was a cluster or a comet, but after further observation, in which he moved locations, the comet's elapsed time movement against the backdrop of stars had proved it was not a cluster but a comet. Seki's report made it into the official domain of discoveries before Lines', so the comet was named Seki-Lines.

By late February and into early March, the comet increased in brightness, now visible to the unaided eye. As it moved through the southern hemisphere constellations of Eridanus and Cetus, the comet continued to brighten further, reaching +4 by the middle of March, attaining −1 by the end of the month.

At a distance of less than 402,000 km from the Sun, and despite the comet's expected brightness to be now exceeding −7 magnitude, it wasn't spotted, only being rediscovered in the evening twilight on April 3, at −2.5 magnitude.

In the evening sky and on view for northern hemisphere observers, Seki-Lines made for an amazing spectacle, with a 15- to

20-degree tail rising straight up out of the western twilight. This was to be its show-stopping moment, as, in the days to come, Seki-Lines began to fade as it passed through Pisces, Aries, and into Taurus, continuing to wane in magnitude all the time.

By April 21, the comet was +4 magnitude and by the month's end was lost the unaided eye. However, it wasn't until early in the following year, January 1963, that the comet was deemed completely lost from the view of Earth.

Comet Ikeya-Seki was a truly magnificent comet, the brightest of the twentieth century, and visible in full daylight within a few degrees of the Sun. Described by some as being 10 times brighter than the full Moon, it was an unforgettable sight for those fortunate to have witnessed it.

Comet Ikeya-Seki

Discovered independently on September 19, 1965, by two Japanese amateur astronomers, Kaoru Ikeya and Tsutomu Seki, it was surprising that the comet had not been detected much earlier, as when it was first sighted the comet was already less than 160 million km from the Sun, and closing in fast.

On October 1, 1965, Kaoru Ikeya and Tsutomu Seki, already in the media spotlight after their discovery, were to be elevated further, after calculations by Dr. Leland E. Cunningham (1904–1989) of Leuschner Observatory, Berkeley, California, predicted the comet's orbit would make it a Sungrazer. Cunningham predicted the comet would pass within 468,319 km of the surface of the Sun, with an expected magnitude of −7 or above. In Japan, where it reached perihelion at noon, the comet was to achieve a magnitude of −10.

The comet was further determined to be a member of the Kreutz Sungrazer family of comets. This orbital stream of 'specialist' comets was discovered by German astronomer Heinrich Carl Friedrich Kreutz (1854–1907) in 1888. Kreutz theorized that Sungrazing comets like Ikeya-Seki all follow the same orbit, further speculating that the comets were most probably debris left over from a much larger, significant comet that had broken up, perhaps due to the gravitational pull it had experienced when passing nearer a larger body.

Kreutz believed that over a period of 800 years, the fragments left by the original larger comet had been further systematically broken up themselves. Occasionally, such a fragment would be drawn or pushed clear from its orbit, causing the body to enter a different route around the Sun, or indeed into it.

The Kreutz Sungrazing comets follow an extremely long, narrow, elliptical orbit, with an orbital period of some 600 years, with the Great Comet of 1106 considered to be the 'parent' body that spawned the debris. During its apparition, the Great Comet was observed globally from the start of February 1106 until the around the middle of March, before it faded from sight. The comet split into two separate fragments, and it is quite possible that one of those fragments became the Great Comet of 1882.

The Kreutz family accounts for around 83% of all Sungrazing comets, with three related groups making up the remainder: the Kracht (Rainer Kracht), Marsden (Brian Marsden), and Meyer (Maik Meyer) groups. The Kracht group and Marsden group have been linked to Comet 96/P Machholz, a short-period comet discovered by amateur astronomer Donald Machholz on May12, 1996.

As for Comet Ikeya-Seki, it wasn't until the last two weeks of October 1965 that interest, which had waned for a time after the initial discovery had been made, was rekindled in both the public's and media's imagination. However, it was to be the last week of October that placed Comet Ikeya-Seki firmly into the Hall of Fame, as in the morning sky observers could see the comet bearing a long and twisted tail that, at its longest, spanned an impressive 113 million km, ranking the comet as the fourth largest ever recorded, with only three previous comets, those of 1680, 1811 and 1843, stretching out longer into space. Comet Ikeya-Seki's tail rose up from the eastern horizon, described as a slender searchlight beam, a truly wondrous sight, as it headed away from the Sun.

However, Comet Ikeya-Seki's nucleus was not to stand the close proximity of its orbit with the Sun. Just before its perihelion passage, the comet broke into three pieces, with the fragments emerging from the other side of the Sun, exactly as they had appeared before they went around it, in almost identical orbits. Comet Ikeya-Seki continued its outward journey with the tail still easily visible deep into November 1965, before it faded completely from view during the opening months of 1966.

Discovered by Gottfried Kirch on November 14, 1680, this comet starred among the brightest comets of the seventeenth century, reported to have been visible during daylight hours and with a substantial tail. Although not thought to have been part of the Kreutz family of comets, the Kirch comet reached perihelion during mid-December of 1680, before attaining its greatest magnitude later in the month. Last sighted on March 19, 1681, credit must also be given to Eusebio Kino, a Spanish Jesuit priest, who, by a detailed study of the comet, charted its course.

Comet Bennett

Formally known as C/1969 Y1, Comet Bennett was discovered in the far southern sky by South African amateur astronomer John ('Jack') Caister Bennett (1914–1990) on December 28, 1969. Comet Bennett reached perihelion on March 20, 1970, passing closest to Earth on March 26 on its outward leg. The comet's movement northwards made for a spectacular object for observers in the northern hemisphere, easily sighted in the east before sunrise.

At its brightest, Comet Bennett achieved a 0 magnitude, with an extensive tail that measured 20°, with the comet's nucleus spouting jets of debris. Last seen on February 27, 1971, Comet Bennett was scheduled to have been photographed by the crew of Apollo 13. Shortly after completing the necessary maneuver to align the spacecraft for a photograph of the comet, the famous mishap took place onboard Apollo 13, and more urgent matters then took precedence.

Comet Kohoutek

Before the arrival of Comet West in 1976, all eyes turned to Comet Kohoutek after its discovery by Lubos Kohoutek, working at the Hamburg Observatory in 1973.

During a search for asteroid images on photographic plates taken in March 1973, Comet Kohoutek was thought to originate from the Oort Cloud. The comet's path showed that this could be Comet Kohoutek's first trek into the inner Solar System and, as a result, could make the encounter potentially a 'comet of the century.' The belief that Comet Kohoutek could be a show-stopper lay in the comet's potentially icy makeup that, on first encounter with the Sun, would make for a spectacular display of outgassing from its nucleus. However, what subsequently transpired was nothing of the sort and, following great disappointment, was to subsequently damage the way in which the media and public viewed comets—until the great spectacle three years later when Comet West arrived.

Before its less than grand entrance, Comet Kohoutek sparked a space and media scramble, with calculations pointing to the comet being large in nature, and with a close passage to the Sun, a strong likelihood of it being a wondrous sight both by day and night. With its perihelion expected to be on December 28, 1973, the world readied itself for the occasion, with NASA launching Operation Kohoutek, a substantial and coordinated program of activities that would encompass and collate observations of the comet from every conceivable viewpoint, including ground-based observatories, aircraft observations and, the ace up its sleeve, Skylab, which could view the comet without the hindrance of Earth's atmosphere.

Heading the project was Stephen P. Maran at the Goddard Space Flight Center, who intended to use the third Skylab mission to watch the comet. With the Russians' Soyuz T-13 also focused on it, the comet was set to make history, being the first to be observed by manned spacecraft, given that Apollo 13's planned observations of Comet Bennett in 1970 were aborted.

Comet Kohoutek promised much and, on balance, did deliver to a certain extent, but not to the grand scale that had been predicted. However, despite the calculations, a lot of the work that went into the expectations surrounding the comet's "great" appearance was, in hindsight, based primarily on speculation and conjecture, leaving a lot to chance.

Even with all the predictions pointing towards such a spectacular encounter, the one element that was overlooked, and that should have been more prominent in the overall reckoning, is that comets are inherently unpredictable, and this taken-for-granted approach to Comet Kohoutek was a sore reminder to just wait and see what happens—prepare as much as possible but never bank on a comet actually being as good as predicted. It is, after all, only a prediction.

Comet Kohoutek's rather close approach to the Sun prior to its rendezvous with Earth was its first undoing. The comet was seen to partially disintegrate and, upon later examination, the previously thought origin of the Oort Cloud had to be revised, as the icy nucleus wasn't as expected. Analysis revealed the nucleus to be more rock-like in formation, pointing to the comet probably having come from the Kuiper Belt instead.

However, Comet Kohoutek did brighten to −3 in magnitude, making for a convincing naked-eye object. The comet also had a distinct tail, reaching 25° in length at best and, like a small number of comets, joined an elite band that also had an anti-tail.

Soon, though, the show was over, and Comet Kohoutek disappeared into the icy depths of space, with its orbit now thought to be a hyperbolic one, meaning it is unlikely that this particular traveler will pass through our region of space again.

Comet West

In 1976, Comet West lit up the late winter/early spring skies and is remembered for being one of the finest comets to grace the inner Solar System (Fig. 3.7).

Discovered photographically by Danish astronomer Richard Martin West at the European Southern Observatory in Chile on August 10, 1975, Comet West attained naked-eye status, reaching perihelion on February 5, 1976, at −3 magnitude. During this time, observers were able to view the comet without optical aid in the broad daylight, making for a wonderful spectacle, once again opening a comet's arrival up to a wider audience.

Comet West gave a fantastic display during its apparition, with a brilliant head and a long, strongly structured tail. At its longest, the dust tail measured a considerable 30° in length.

Fig. 3.7 Comet West (Credit J. Linder and ESO)

On March 7, 1976, it was reported that the nucleus of Comet West had fractured into two pieces, with further reports, in particular from Steven O'Meara on March 18, using the 9-inch Harvard refractor, suggesting that the comet had further split, now displaying four individual parts.

According to some observers, no less than five components existed, with the close approach to the Sun having had a damaging influence on the comet's nucleus, literally tearing it apart as the Sun's intense heat caused it to explode. What was left of the comet after its encounter with the Sun is unlikely to be seen anytime soon, as the comet is thought to have a near-parabolic orbit. However, despite such an eccentric orbit, it is of significant interest to note the history of visits made by the Comet West before the fateful visit of 1976.

With an estimated orbital period of 558,000 years, the return of 1976 was about the 120th time the comet had returned since the extinction of the dinosaurs, and the 8000th time it had returned since Earth was in its early stages of formation, and under bombardment from a swarm of comets. It is quite possible that Comet West 'witnessed' the arrival of Homo sapiens on Earth. One can but speculate as to what Comet West will find upon its next visit should it ever return, but, judging by the substantial passage of time during its absence, the colonization of other planets would seem to be well within the realms of possibility, and even spaceflight to the nearest stars, where inhabitable planets exist.

Comet IRAS-Araki-Alcock

Formally designated C/1983 H1 and known previous to that as 1983 VII, Comet IRAS-Araki-Alcock adorned our skies during 1983. Its host of discoverers made for quite an assortment from the world of astronomy. They comprised the Infrared Astronomical Satellite (IRAS) and amateur astronomers George Alcock (1912–2000), based in the UK, and Genichi Araki of Japan.

First sighted by IRAS on April 25, 1983, the media was slow to pick up on the discovery until it was 'humanized' on May 3, when Alcock and Araki announced their own individual observations of a new comet. It is also worth noting that astronomers working at Uppsala University in Sweden had also filed observations of a new comet around the same time, although, for the most part, nobody paid them much attention.

Comet IRAS-Araki-Alcock was the seventh comet to be discovered during 1983, making a very close approach to Earth during its visit. At a distance of just 4.6 million km, it was the closest any comet had been to Earth during the past two centuries, with the exception of Comet Lexell in 1770.

Comet Lexell was discovered by Charles Messier on June 14, 1770, but was known as Lexell after its orbit was determined by Swedish astronomer Anders Johan Lexell (1740–1784). On July 1, 1770, the comet passed Earth at just 2.25 million km, approximately six times the radius of the Moon's orbit. It was noted at the time by an English astronomer that the nucleus of Comet Lexell was as large as Jupiter, "surrounded with a coma of silver light, the brightest part of which was as large as the Moon's orb."

According to Lexell, the comet followed an elliptical orbit rather than the suggested parabolic orbit. At the time, it was largely considered that all comets originated from outside of the Solar System, and that if calculations were to be made to attempt to establish an orbital time frame, it must be based on a parabolic orbit.

However, Lexell was correct—not only with the elliptical orbit, but by further stating that, because of its interaction with the gravitational forces of Jupiter in 1767, it is unlikely to have been previously seen and documented. Furthermore, Lexell had, by his calculations, discovered the first near-Earth object.

Lexell had calculated that the orbit of the comet was elliptical, with a 5.6-year orbital period that took in around the Sun and then out towards the region of Jupiter. However, it seems to be the constant dalliance with Jupiter that was to ultimately swing Comet Lexell into a completely different orbit following its no-show in 1776. Effectively, Jupiter had altered the comet's orbit so substantially that it was to never to be seen again. Comet Lexell had effectively been ejected out of the Solar System.

By comparison to Comet Lexell, Comet IRAS-Araki-Alock is a long-period comet of around 964 years, and the parent comet of a minor meteor shower, the Eta Lyrids. During its closest approach to Earth in 1983, the comet appeared as a hazy circular cloud, about the size of a full Moon, although because we were seeing the comet head-on, it showed little tail of any note. However, the comet brightened to naked-eye magnitude or 3–4, moving swiftly across the sky at 30° a day, increasing to 60° across the sky at closest approach. It was so close that the brightness of Comet IRAS-Araki-Alock was diffused across the sky, making it appear strangely faint.

Comet Hyakutake

Dubbed the Great Comet of 1996, Comet Hyakutake proved to be most enlightening on several levels. It was discovered on January 31, 1996, by Yuji Hyakutake (1950–2002), an amateur astronomer from southern Japan who had already made a cometary discovery several weeks earlier. The comet at this point had a +11 magnitude and, during the course of the coming months, was to pass Earth at one of the closest ever recorded distances.

A long-period comet, Comet Hyakutake had thought to have an orbital period of around 17,000 years, but, during its 1996 passage through the Solar System, gravitational perturbations of the larger planets were thought to have significantly increased this orbit from 17,000 to possibly as much as 70,000 years.

Yuji Hyakutake's discover gives much hope to comet hunters who rely solely on binoculars to make their finds. Hyakutake, using a powerful set of binoculars (25 × 150 mm), not only discovered his first comet with them but also his second, the more famous comet. The first discovery, made on Christmas Day 1995, never made it to naked-eye brightness, but upon returning to survey some five weeks later the area where he had found the first comet, a second comet was discovered. The second comet was almost in the same position as the first had been, with his observations confirmed by independent observers over the hours to follow.

Comet Hyakutake was to attain naked-eye brightness, staying a naked-eye object for three months, making it the brightest comet seen for 20 years. The comet could be observed overhead during pre-dawn hours, reaching 0 magnitude in brightness. At its longest and most magnificent, Hyakutake graced the skies with a tail boasting in excess of 100° in length. Hyakutake was to pass very close to Earth, some 14.9 million km, the intrinsically brightest comet to pass so close since 1556.

The Great Comet of 1556, sometimes referred to as the Comet of Charles V, was recorded to have an apparent diameter equal to half that of the Moon, and is further thought to be the same comet as the Great Comet of 1264, one of the brightest comets on record, visible from July until September of 1264.

Clearly, having such scientifically advanced craft in space during the time of any approaching body, be it asteroid or comet, can serve as an extraordinary tool in continuing to unlock the mysteries of the universe. Often by chance, but more by design in the modern era, our understanding has leaped tenfold using such spacecraft. Outside of our own atmosphere, which both restricts and protects, it is a different world, as was to be proven by the study in space of Hyakutake.

One such non-intended encounter occurred during the visit of Hyakutake, when the Ulysses spacecraft, primarily put into space to orbit and observe the Sun, passed through the tail of the comet. This was not the only freak encounter that Ulysses was to stumble across, when later it passed through the tail of Comet McNaught in 2007.

Launched back in October 1990 aboard the space shuttle *Discovery*, having previously been scheduled for launch on *Challenger* until its tragic loss, Ulysses—using most notably its SWOOPS (Solar Wind Observations Over the Poles of the Sun) instrument and magnetometer—registered some very bizarre readings on May 1, 1996. The solar wind, which should have been present, was registering to a greatly reduced degree, being replaced by gases not normally found in the solar wind. Also, the magnetic field within the solar wind was noticeably distorted.

It transpired that Ulysses, despite its great distance from Hyakutake, had actually wandered through the tail of the comet, measured by the craft as a staggering 579 million km in length. This made Hyakutake's tail the longest ever recorded, breaking the previous claimed record of 331 million km set by the Great Comet of 1843. Although there is some dispute, Hyakutake's tail is the longest that we have been able to measure.

Hyakutake also had more surprises in store, with large quantities of ethane and methane discovered, molecules not previously known to exist in comets or in interstellar matter. The discovery also suggested that Hyakutake could possibly be a new type of comet and that, consequently, at least two basic types of comet inhabit our Solar System. The composition also pointed towards Hyakutake's birthplace surroundings being very much different from its counterparts that do not contain ethane.

The presence of ethane in such a body had a profound impact on the scientific theories that describe the primordial conditions that led to the formation of the Sun and the planets. Levels of ethane in Hayakutake were about 1000 times greater than can be explained, if the molecules were formed by normal physical processes within the gases of the primordial solar nebula, the very birth cloud of the Solar System.

Murchison Meteorite

On the morning of September 28, 1969, near the town of Murchison, Victoria, in Australia, a bright fireball streaked across the sky. As it did so, it was observed breaking into three separate fragments, leaving a trail of smoke in its wake, with a noise likened to that of a tremor heard less than a minute later.

The fragments, strewn over an area possibly as wide as 16 km^2, were in time collected, with a total collected mass of 100 kg. Like Comet Hyakutake to come, the Murchison meteorite also revealed the presence of ethane and methane, perhaps leading to the conclusion that both the comet and meteorite were of common origin, born around the same time, in conditions totally at odds with other comets and meteorites that had previously been studied.

Hale-Bopp

Discovered independently on July 23, 1995, by two American observers, Alan Hale and Thomas Bopp, Comet C/1995 01 was to become probably the most widely observed comet of the twentieth century and one of the brightest visitors to adorn our skies for many a decade. Comet Hale-Bopp subsequently became entrenched in the annals of history as the comet that was visible to the naked eye for a record 18 months.

In the case of Alan Hale, it was the reward for many hundreds of hours searching for comets without success, but, such was the determination and dedication, his efforts were to reap substantial rewards following his discovery, as the media bandwagon gathered momentum. At the complete opposite end of the spectrum to Hale, Thomas Bopp had chanced across the comet while looking through the eyepiece of a friend's telescope!

When sighted, the comet had a 10.5 apparent magnitude, spotted near to M70, a globular cluster in the constellation of Sagittarius. Both Hale and Bopp, upon discovering the object, consulted their star charts to rule out any other possible phenomenon that could account for what they had seen, but nothing was cataloged. Still working independently and unaware of the other's movements, Hale and Bopp then both contacted the Central Bureau for Astronomical Telegrams, the clearinghouse for astronomical discoveries. Following subsequent checks, the morning brought forth news that the pair had found a new comet.

Their discovery was already record-breaking, logged as the farthest comet from the Sun to be discovered by amateurs, indeed, another landmark registered by those outside of the scientific community, testimony to exactly what can be achieved with involvement at that level. First sighted when the comet was beyond the orbit of Jupiter, this made for a very special occasion; it was literally a thousand times brighter than Comet Halley at the same distance.

It became established that Comet Hale-Bopp was most likely an Oort Cloud object, heralding from the outer Solar System, with an estimated age equivalent to the Sun: 4.5 billion years.

With its longevity, it is quite possible that the comet had been seen and documented much earlier in history. It is thought that Hale-Bopp may well have been observed by the ancient Egyptians during the reign of Pepi I (2332–2283 BC). In Pepi's pyramid in Saqqara, an area south of modern-day Cairo, there is text referring to an "nhh-star" as a companion of the pharaoh in the heavens, where "nhh" is the hieroglyph for long hair. With an orbital period of 2392 years, it is more than likely that other sighting records exist but haven't widely come to light.

Comet Hale-Bopp's brightness was attributable to the size of its nucleus, measuring some 30–40 km in diameter. Considering the average comet has a nucleus that measures around 5 km, this was a giant, probably even larger than the body that caused the extinction of the dinosaurs, with some estimates putting the size of that nucleus at 10–14 km.

During the comet's apparition, it was visible to the naked eye between May 1996 and November 1997, registering a magnitude greater than 0 for eight weeks, longer than any other comet. Perihelion occurred on April 1, 1997, with the comet's closest approach to Earth being measured at some 193 million km. Visible in the northeastern sky at dawn, and northwestern sky at dusk, its twin blue-and-white tails both measuring 15°–20° long were easily observable, making for a wonderful early and late spectacle in our skies, the like of which may not be seen again for some time. The appearance of Comet Hale-Bopp does have a tragic side to it, with about 39 members of the "Heaven's Gate" cult committing mass suicide in San Diego at that time.

Comet Hale-Bopp will most likely make a scheduled return, although estimates vary wildly as to when, because of the close interaction with the gravity fields around Jupiter. It is thought that its original orbit of 4200 years—a highly elongated orbit that results in the comet going beyond Pluto—has been shortened to around half, something in the region of 2300–2500 years. Therefore, there is no definite due date for the comet's return, and, with so many large bodies known and unknown that can alter cometary orbits, no comet that makes such a trek as Comet Hale-Bopp can be excluded from the possibility of such orbital deviations.

Comet McNaught

Labeled the Great Comet of 1997, Comet McNaught, originally designated C/2006 P1, dazzled observers with its brightness and spectacular tail, although its apparition was to only delight those in the southern hemisphere, with a much downgraded display offered to those in more northerly climes.

Discovered on August 7, 2006, by prolific asteroid and comet hunter Robert H. McNaught, the comet was to reach a peak brightness of −5.5 magnitude, the brightest seen in our skies since 1935. While working at the Uppsala Southern Schmidt Telescope, McNaught, a British-Australian astronomer, made the discovery during part of a project known as the Siding Spring Survey, undertaken to study near-Earth objects (NEOs), until funding dried up in 2013. Shining very dimly at a brightness of +17 in the con-stellation of Ophiuchus, McNaught came across the comet on a CCD (charged coupled device), image while making routine observations for the Siding Spring Survey.

With over 50 comet and asteroid finds to his name, McNaught tracked the comet as it moved through the constella-tions of Ophiuchus and Scorpius, brightening from +17 to +9 as it sped across the heavens. For a time the comet was lost, only to be sighted again steadily brightening in the constellation of Sagittarius in January 2007, going on to reach naked-eye visibility.

On January 12, Comet McNaught reached perihelion, passing around the Sun but still being watched as it did so by the Solar and Heliospheric Observatory (SOHO). SOHO, a joint ESA and NASA project, was launched on December 2, 1995, on a two-year mission to study the Sun, although the actual functioning of the craft went far beyond the intended life of the spacecraft.

Having rounded the Sun, Comet McNaught then made for a wonderful spectacle in the southern hemisphere sky just after sunset. The comet attained its greatest brightness of −5.5 magnitude just before perihelion, emerging to still shine brightly from the other side of the Sun at −4 magnitude. The comet was to adorn southern skies in a glittering display, visible to the naked eye even in broad daylight, with a magnificent fan-shaped tail.

Having passed through the tail of Comet Hyakutake in 1996, the Ulysses spacecraft was to make yet another unscheduled encounter, this time with Comet McNaught, and where before the passage through Comet Hyakutake's tail took under three days, it was 18 days before Ulysses passed through the area of solar wind affected by Comet McNaught. Measuring the equivalent of about 35° long, around the same apparent size as 70 full Moons lined up in the night sky, the tail appeared enormous, with a number of luminous bands adding to its glory as it swept around in an arc from McNaught's core, which itself measured 24 km in diameter.

Comet Lovejoy

Using a 7.9-inch Schmidt-Cassegrain telescope from his backyard in Queensland, Australia, amateur astronomer Terry Lovejoy added to his cometary finds with a third new find that was to be nicknamed "The Great Christmas Comet of 2011." Discovered on November 27, 2011, C/2011 W3 Lovejoy is one of the Kreutz Sungrazer family of comets, and during its last apparition, lived up to the Sungrazer reputation by making a remarkably close circuit of the Sun, at a distance of just 140,000 km from its surface, taking it through the Sun's corona.

A long-period comet, Comet Lovejoy was first sighted in Puppis during a comet survey, reported by its discoverer as "a rapidly moving fuzzy object" of +13 magnitude. Less than a week later, the comet became visible to the STEREO-A (Solar Terrestrial Relations Observatory) spacecraft, which, together with its companion STEREO-B, was launched on October 26, 2006. The purpose of the mission was to provide the first-ever stereoscopic measurements of the Sun and related space weather, including observations of coronal mass ejections. By mid-December, no fewer than six satellites including the SOHO were watching Comet Lovejoy, with instruments on board SOHO detecting a small companion fragment traveling alongside the main body.

Comet Lovejoy reached perihelion on December 16, 2011, literally grazing the Sun and, in doing so, lost a chunk of its 500-m-wide nucleus. The loss of nucleus mass may well account for its noticeable dimming after rounding the Sun, but having 'gathered' itself somewhat, the comet still put on a fine display, sporting a 40-degree tail that was to extend upward from the southeastern horizon when viewed from the southern hemisphere. Having attained a brightest magnitude of between −3 and −4 for a proportion of observers, the comet's close proximity to the Sun meant for tricky if not impossible viewing.

By the start of 2012 Comet Lovejoy had all but faded, soon to disappear from our skies. With an estimated orbit of 622 years, and with such an eccentric path and inclination that avoids any significant interaction with the gravitational fields surrounding the larger outer planets, Comet Lovejoy may well put in another appearance, with an estimated return in the year 2633.

The Oort Cloud

As one cometary mystery was unraveled, a new puzzle was presented. With the advancement in technology that allowed orbits to be more precisely determined, it became apparent that most comets followed elliptical orbits, and were therefore considered members of our Solar System. But what about the orbits of long-period comets, where bodies weren't confined to orbits within our Solar System, venturing inward from interstellar space?

Beyond Neptune, at nearly halfway to the nearest star, lies the Oort Cloud. Surrounding our Solar System, the cloud is literally a vast reservoir of comets, an immense spherically shaped cloud held in place at the most extreme reaches of the Sun's gravitational capability. The existence of the cloud was first inferred by Dutch astronomer Jan H. Oort (1900–1992) in 1950.

Using the physical evidence of long-period comets entering the planetary system, Oort believed a field of cometary bodies must exist in our own Solar System. He proposed that these comets remained within this cloud until the influence of a passing star diverted them towards the inner part of our Solar System. Oort, who determined the rotation of the Milky Way Galaxy during the 1920s, further proposed that, once the comets had left the cloud, the Sun's influence would capture them gravitationally, stabilizing them into an orbit that could last thousands of years. Oort, using data collected on the orbits of 19 comets, came to the conclusion that they all must have come from the cloud, with later studies conclusively confirming that what Oort had envisaged was correct, and responsible for many of the comets cataloged over the centuries.

It is not just long-term comets that have been attributed to the Oort Cloud. Halley's Comet was thought to originally be a higher-inclination intermediate comet, pulled in from its erratic orbit into a shorter one, not just by the influence of the Sun but also by the major planets, which are responsible for the shaping of many comet orbits.

However, the Oort Cloud does have its doubters, and as science continues to expand its knowledge and therefore challenge the accepted, the Oort Cloud, while competently explaining a lot of what is known, is, in the view of some scientists, fundamentally flawed. If the Oort Cloud was to be found wanting for the conclusive answer as to where many comets come from, a serious re-examination of the celestial mechanics of comets would need to be undertaken, with a domino effect on theories of mass extinctions that could well have shaped life on Earth.

Although the Sun's gravity holds the Oort Cloud and its contents in place, the hold is weak, which is why influences from other larger bodies can disturb the pattern and distribution of the comets within the cloud. The outer boundaries of the Oort Cloud are thought to be more susceptible to influence, as the Sun's gravitational reach continues to weaken at an increased distance. The bulk and stability of the cloud is thought to be 'compacted' near to the ecliptic plane, where bodies on the outer limb are at intervals pushed inwards from the outer part of the cloud as the result of external forces from larger passing bodies. This gives the denser region of the cloud a constant feed to maintain its mass. From the Sun, and more significantly from stellar perturbations, the cloud is constantly being influenced, with matter pooled from a vast array of sources accounting for the different structures of the comets that herald from within it.

The Oort Cloud and Life Extinction on Earth

The Oort Cloud, for the most part, appears to be in a reasonably stable region of space located at the very edge of our Solar System. However, at a lower percentage of times, this can become a dramatically unstable region of space.

The Oort Cloud is occasionally responsible for producing comets that may fall into a regular path around the Sun, but every 26 million years, during a phase of major instability, could the cloud possibly be the source for mass extinction events on Earth? What if something were to create such a dramatic influence on the Oort Cloud that a vast number of comets were released all at one time, a huge venting of bodies that, quite possibly, were to head in towards the inner Solar System, and ultimately, towards us?

Could the Oort Cloud be responsible for a periodic bombardment of projectiles towards Earth, leading to an extinction-level event every 26 million years? During this time frame, fossil records constantly point to an abnormally high number of species being wiped out, an event that can be traced back 250 million years, including the period consistent with the extinction of the dinosaurs at the end of the Cretaceous period.

What then would influence the Oort Cloud to produce such a mass migration of comets from within it, a veritable storm of comets? One such explanation lies with the periodic oscillation of our Solar System, a process whereby there is the motion of moving back and forth across the plane of the Milky Way, this movement occurring in the region of every 30–35 million years. Extinctions due to the cosmic bombardment of debris (setting aside the argument that a large individual asteroid was responsible) would be triggered when our Solar System was crossing the densest part of the galactic disk, the Oort Cloud being severely imposed upon during this time. The subsequent influence placed upon the cloud would cause it to vent huge numbers of comets at once.

Another theory has suggested the presence of giant molecular clouds situated in the mid-plane of the galactic disc, these clouds intermittently influencing the Oort Cloud. However, the 'disk tides' seem to be the strongest candidate in answering the question. These disc tides are the cumulative effects of local matter in the plane, perpendicular to the galactic disk.

The disc tides theory seemed to satisfy much of the debate, but it remains unclear as to how their impact would modulate the flux of comets at different heights above or below the galactic mid-plane. For many years following the disc tide explanation, it was generally considered in the astronomical community that this was the more likely answer, although the complete mechanics remain unsolved.

Later research, rather than picking apart the theory because it didn't offer a total solution, in fact reinforced it. In 1995, John Matese and Patrick Whitman of the University of Southwestern Carolina, along with Mauri Valtonen of Finland and Kimmo Innanen of Canada, injected new positives into the effects of these so-called disk tides. Their numerical models of the Oort Cloud mechanics suggested that as our Solar System oscillates through the galactic plane, the tides modulate the release of comets from the cloud at a ratio of 4:1, with the greatest effect taking place at mid-plane. This would explain the copious venting on the timescale that had been originally suggested.

John Matese continued his work in the field and, with colleague Daniel Whitmire, took the theory a step further. After studying a selected group of comet orbits, the two proposed that the entire galaxy, including the distant matter at its central core, plays a role in hustling some comets free of the Oort Cloud. Unlike the influence of disc tides, these distant-matter tides exert themselves within the plane of the galactic disk, accounting for as much as nearly one-third of all comets that we observe, with around two-thirds the result of disc tides. The remaining cometary releases are the result of the influence of nearby stars and, as suggested in an earlier theory, the presence of giant molecular clouds.

Paul Weissman of the Jet Propulsion Laboratory in Pasadena, California, was one who remained unconvinced. He has ultimately concluded that asteroids, which account for 75% of impact craters on Earth, play a greater role in impact extinctions.

4. Meteors and Meteorites

For some observers, interest in astronomy was first sparked by the sight of a meteor blazing its way across the heavens. For some of the curious, the spark subsequently turned to a flame, as the desire to find out what they had seen soon followed. The wish to see more and in abundance would fan the flames, until the desire and passion for astronomy itself would take hold.

Meteorites have been recorded throughout history, with meteoritic iron being used to make tools as early as 4000 BC, and the oldest known artefact found in northern Egypt being dated at 3200 BC. This artefact consists of nine small beads, hammered into shape by being first flattened and then rolled. Despite the chemical content of the beads being disputed, the argument was settled through the use of microscopy, and the discovery of Widmanstätten patterns.

Many civilizations across the globe have made accounts of meteorite findings, some worshipping the stone that had been discovered, believing it to have sacred powers. Early Japanese records show meteorites being regarded as religious artefacts.

Egyptian hieroglyphics on the walls of pyramids referred to meteorite finds as "heavenly irons," with the recovery of knives and weapons from tombs of Egyptian pharaohs attesting to the iron's use. To the Egyptians, the sky was of great importance, and any object that fell from it would be considered sent by the gods. Archaeology on Aztec, Mayan and Inca civilizations have also revealed the manufacturing of weapons from meteorite finds.

North American tribes of the Kiowa regarded the Willamette meteorite, a 15.5-ton colossus, as their central sanctuary. Discovered in Oregon, the Willamette meteorite is the largest find in America, ranking sixth largest in the world. Before a tribe embarked on hunting expeditions, the men dipped their spears and arrow tips into the water that had gathered in the large hollows of the iron meteorite. They believed that, in doing so, spear and arrow alike would be made accurate and fast, just like a meteorite.

© Springer International Publishing AG 2017
J. Powell, *Cosmic Debris*, Astronomers' Universe,
DOI 10.1007/978-3-319-51016-3_4

In the seventeenth and eighteenth centuries meteorites were also known as 'thunderstones,' most likely because of the associated sonic boom that fireballs can produce as they travel through the atmosphere.

Classification of Meteorites

Meteorites are classified into three basic categories: iron (siderites); stony (aerolites); and stony-iron (siderolites).

Iron

Iron meteorites generally herald from the Asteroid Belt. Their composition suggests that they were once part of the core of a planet or asteroid. They are composed of iron alloyed with nickel. Iron is the predominant metal, combined with a lesser amount of nickel, with a trace of cobalt also showing. However, the ratio is significantly in favor of iron, perhaps more than 90% (Fig. 4.1).

Iron meteorites generally feel weighty when compared to standard Earth rocks of the same size and, because of their content, are easily attracted to strong magnets. There are three classes of structural makeup within the iron classification.

The first class is hexahedrites, which have a hexagonal (six-sided) crystal shape. Hexahedrites contain a very high percentage of the nickel-iron alloy kamacite. To a much lesser extent, in some cases, there are trace elements such as carbon, chromium, cobalt, phosphorus, silicon, and sulfur.

When they are cut and polished, hexahedrites show interesting configurations within them known as Neumann lines or bands, produced by compressional shockwaves. Named after their discoverer, German mineralogist and mathematical physicist Franz Ernest Neumann (1798–1895), these patterns are aligned along the faces of the cube. Although most obvious in hexahedrites, Neumann lines are present in other meteorites.

Octahedrites, whose crystalline formation parallels that of an octahedron, and ataxites make up the other two structures, although

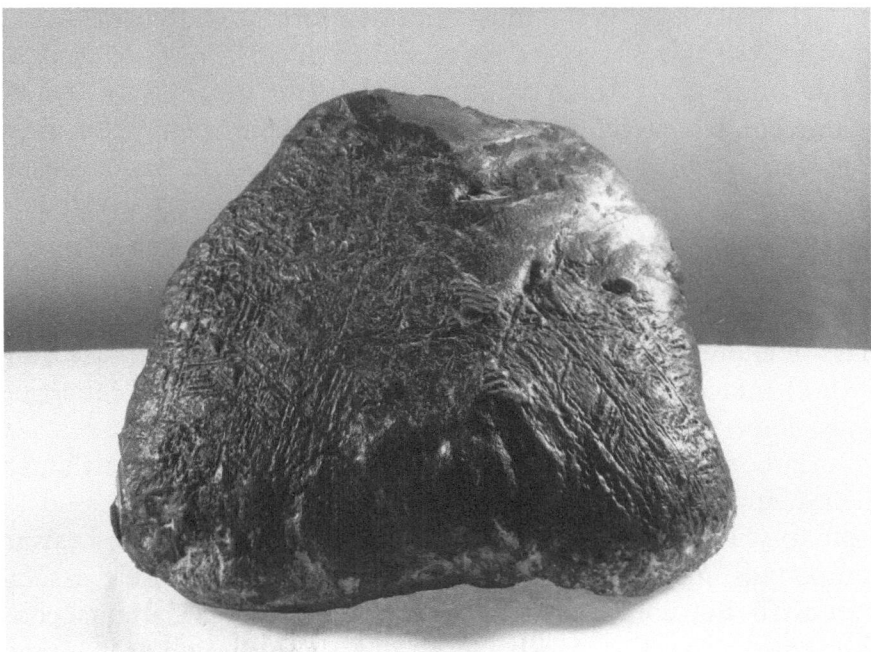

Fig. 4.1 Iron-nickel meteorite found near Fort Stockton, Texas, 1952. Measuring 15 centimeters across, the meteorite displays the familiar triangular Widmanstatten pattern, thought to have formed more than 4.5 billion years ago as the metal cooled (courtesy of NASA)

the word ataxite does herald from Greek, meaning "without struc-ture." Although octahedrites when cleaned and acid etched show Widmanstätten patterns, ataxites show no pattern on etching.

Stony

This is the largest known classification of meteorites, thought to have once formed the outer crust of a planet or asteroid. A proportion of stone meteorites contain small, grain-like inclu-sions known as chondrules. This forms a sub-divisional class of meteorite, with stone meteorites containing these chondrules known as chondrites. Meteorites that do not contain any chon-drules are known as achondrites. Their origin is very different from that of chondrites, more likely to be of a volcanic nature, whereas chondrules are formed in a solar nebula.

The chondrite is the most common form of meteorite, taken from the Greek word *chondros,* meaning "seed." They are thought to have formed at around the same time as the inner rocky planets of our Solar System. These chondrites, despite being the most common, are of great interest, as here we have a hands-on object from a crucial time in our past. The chondrite offers a valuable insight into the early mechanics and formation of Earth.

Chondrites are divided into five main classes: ordinary (O); carbonaceous (C); enstatite (E); kakangri (K); and rumuriti (R). Each class of chondrite was formed at different distances from the Sun, with O meteorites accounting for the vast majority of cataloged finds, some 90%.

Carbonaceous chondrites account for many fewer, although their composition is far more intriguing than those classed as ordinary. These carbonaceous chondrites have carbon-bearing compounds and, overall, it is these that are more closely associated with the composition of the inner planets. Carbonaceous chondrites also tell us much about the early formation of the inner planets, where the initial particles began to form a mass over time into a planet-sized body. These chondrules, typically a millimeter in size, were formed by the heating and then bonding together of dust grains.

These grains, when acting on a much larger scale, attracted other material, probably organic, and this formed a kind of gelling agent, eventually forming planetesimals. In turn, the planetesimals then commanded their own gravitational field and, with gravity in place, the collective binding force would grow increasingly stronger, until the body evolved into a planet.

Enstatite chondrites have very oxygen-poor compositions, showing less oxygen for bonding than other meteorite groups. Kakangri have a dust-like matrix, not dissimilar from carbonaceous chondrites. Rumuruti chondrites, while also containing a dusty matrix, are far more oxidized in nature, virtually absent of metal.

Chondrites are further classified from 1 to 6, based on the amount of heat they have been exposed to, with 1 representing unheated and 6 showing exposure of the meteorite to extreme heat.

When examining our own terrestrial-based rock against that of a chondrite, the one feature of the composition that clearly defines the difference is that the chondrite almost always contains minute fragments of metal (iron and nickel), which is why many hunters use a special form of magnet when searching for meteorites. The magnet's composition makes not only the fragment more detectable but also proves to be a good test that what is found is actually is a meteorite.

The presence of metal also determines that the rock in which it was found has not been melted since the formation of the Solar System. The metal within the rock has not sunk to the core of its parent body, typically an asteroid, unlike what has occurred with Earth's core.

Inside what can be a rather complex diversity of composition within a chondrite meteorite, we find traces of diamond, silicon carbide, corundum, silicon nitride and graphite. This would indicate that certain types of chondrite were present before the formation of our Solar System, with them possibly dating back to the formation of stars.

Stony-Iron

Accounting for the smallest proportion of the main three classes are stony-iron meteorites, with their formation believed to have occurred at the core/mantle boundary point within the parent body. These stony-iron meteorites are generally composed of equal amounts of nickel-iron and stone. In turn, these are subdivided into two groups: pallasites and mesosiderites.

Pallasites are made up of a nickel-iron matrix that includes tiny olivine mineral crystals of peridot quality (gem-like) and, if revealed in sufficient quantity, display an emerald-green appearance. Also, in the case of some pallasites, when cut, so-called Widmanstätten or Thomson patterns are revealed. Named after Count Alois von Beckh Widmanstätten (1753–1849), these ribbon-like features consist of a delicate interweaving of kamacite and taenite, alloys of iron and nickel.

Despite the apparent accolade for the find, which was credited to Widmanstätten, there is a counter and possibly more credible claim for the discovery from Guglielmo Thomson (1760–1806). Thomson, an English geologist, used nitric acid to clean up a portion of the Krasnojarsk meteorite and, upon doing so, noticed a strange and hitherto undiscovered appearance to the metal from which he was attempting to remove rust.

The first pallasite meteorite was found on the side of Mount Bolshoi Imir, around 193 km southeast of Krasnojarsk, Russia. The discovery was made by Yakov Medvedev (of whom little is known), who at the time was searching for ore veins. During his search, Medvedev had stumbled across a large iron boulder, but it wasn't until over two decades later that, having seen the boulder for himself, a German copper miner named Johann Kaspar Mettich decided to write a report to renowned explorer and fellow German Peter Simon Pallas (1741–1811). Pallas, also a zoologist and botanist, saw for himself this 680-kg lump of metal and duly instructed it to be sent to St. Petersburg for analysis.

Pallas, having recognized the potential uniqueness of the find, encouraged Ernst Florens Chladni (1756–1827), a German physicist known as the father of meteoritics (the study of meteorites), to make a study of the find, in part, to convince and ultimately silence a number of skeptics. Analysis proved that the 'rock' was not of Earth but from space. Pallas, for his actions, and for the subsequent discovery that the Krasnojarsk meteorite was a different type of meteorite, had the classification named after him.

Mesosiderites, derived from the Greek words for 'half' and 'iron,' is the smaller and much rarer subdivision of the two stony-iron groups. They contain both nickel-iron and silicates. When cut and polished, they reveal a silver and black matrix (Fig. 4.2).

Meteoroids

The origin of the meteorite is linked to meteoroids, larger bodies that come into contact with Earth. Unlike the smaller meteors, which burn up in the atmosphere, debris from the breakup of the larger meteoroid can make it to the surface. A typical meteoroid

Fig. 4.2 The etched surface of a fragment of the 3.9 billion year old Mundrebilla meteorite, one of the largest meteorites ever found. Discovered in Western Australia in 1911, the collective weight of the rock is estimated at 24 metric tons (courtesy of NASA)

can be less than 100 kg in weight but, accompanied by its larger size when compared to the dust particle or pebble-sized meteor, has much greater potential of making it to Earth's surface. In one year, it is estimated that perhaps as many as 17,000 meteorites fall to Earth, many larger than 100 g.

The meteoroid, traveling through space at speeds up to 70,000 mph (112,500 km/h), will find its path slowed as it enters Earth's atmosphere, with the frictional forces imposed on the body acting on its extremities and, in doing so, heating the object to a temperature that makes the meteoroid begin to melt. As the melting process accelerates, bearing in mind the incredibly short time frame for this occurrence to take place, droplets of molten material begin to break free from the mass of the body. Whereas vaporization would have by now burned smaller meteorites to a fiery death, the meteoroid plows on, continually slowing as it does so, with the heating replaced by a cooling effect.

The result of the rapid cooling coats the meteoroid in a glassy-type shell, a process called fusion crust, and from here, as its descent nears the ground, gravity takes over, bringing the lump of rock down to Earth with a literal thud. During its passage through Earth's atmosphere, and given its speed and the resulting friction, shockwaves are generated, so not only can observers for many surrounding kilometers see the meteoroid, they can hear it as well, courtesy of the accompanying sonic boom. As it rests at the end of its journey on Earth's surface, having potentially produced a crater, its classification turns from meteoroid to meteorite.

Because of its passage through the atmosphere, meteorites may have a distinct shape to them, curved on one side where the facing edge of the body has encountered resistance on its passage, friction molding its appearance. The curvature can be even more elaborate, perhaps more conical in shape, with rate and angle of descent determining just how acute the final curve is.

The resulting thud of impact is unlikely to cause further augmentation to the final shape, as what meteorite hunters normally find is the finished article, rather than an object whose look has been determined by impact. It is only the outermost skin of the meteoroid that has melted, its interior remaining cold and hard— exactly how it was before entry into Earth's atmosphere. Those who recover the meteorite generally find the rock in a cold state. However, if spotted and retrieved early enough, the object can possibly be warm to the touch, but nothing excessive. For those meteoroids with less of an easy passage—those that rotate, spin, or tumble—a more irregular shape can be presented, although, even here, as the outer skin is intermittently revolved to be the leading edge, the object can still be of a semi-smooth nature.

With 70% of Earth's surface covered in water, many meteorites simply go unrecorded and undiscovered; others fall on barren wastelands, subjected to many years of weathering. However, as the composition of the body has a percentage of metal within it, this can act as a preservative, slowing down the erosion. A desert-like landscape or vast plain can offer ideal hunting ground for the meteorite enthusiast, with many being discovered and subsequently cataloged, making these rocks known as 'find' meteorites. 'Find' meteorites are generally iron in composition, while the closely associated 'fall' meteorites—those witnessed following perhaps a fireball—are dominated by the stonier kind.

Using observational data on meteors and fireball sightings, it has been determined that meteorites fall evenly across the globe, although falls in the polar regions, while cataloged, are less likely, as the parent meteoroid is more likely to be traveling on the same plane as other bodies in our Solar System, the planets. Although this is not a hard and fast rule, it does seem to dictate and correspond with fall patterns.

The data calculated from sightings also implies that the majority of falls occur in the afternoon and evening. This is because as Earth rotates the hemisphere that presents itself with noon to midnight local time encounters bodies that are in orbit around the Sun and closing in on Earth, these objects subsequently being drawn towards Earth by the planet's gravity. Objects that pass in front of the hemisphere with local time between midnight and noon are pulling away from Earth, with gravity less likely to have any effect on the object.

Many of us will never find or let alone see a meteorite fall, but should you witness one, or have sufficient grounds to believe you've found one, the find should always be reported to the local astronomical society, college or university. Details such as time of sighting, or time of find, and any accompanying information such as the direction the meteorite seemed to come from and co-ordinates of the find are invaluable. Not only is it a rarity to witness a fall, but the data provided can help and assist in the further understanding of these bodies.

Meteorite Origins

Our Moon

Meteorites arrive on Earth from three distinct sources. The vast number of them, 97%, herald from asteroids, and not necessarily those in the Asteroid Belt. To a much lesser extent, a few meteorites have been cataloged with our own Moon documented as their origin, with fewer still heralding from the planet Mars. A possible fourth source of meteorites is comets, although this is highly debatable. It would seem plausible that a comet could eject such debris through a collisional demise, and the fragments could be classed as a meteoroid, which could then enter Earth's domain. The argument loses credibility with compositional issues seemingly bringing all speculation to a halt. Our ability to go the Moon has given the chance to actually visit one of the places from which some of the meteorites have actually come, thus affording us the ability to par-up findings that, for most, remain impossible to prove.

It is easy to see in the night sky how much the Moon has had to endure in its past, not just from its violent volcanic era but from meteorite strikes and from bodies of a larger size that have left much of the lunar surface pitted. The one thing the Moon does have in its favor is that it makes for a relatively small target, but, as the dinosaurs discovered, that does not mean that smaller bodies are exempt from possible life-changing impacts.

In 1981, a meteorite was discovered during an American-led expedition to Antarctica. The find, Allan Hills (ALH) 81005, was dark gray in color, with centimeter-sized patches of a white material. Subsequent investigation, when compared to lunar rock samples brought back from the Apollo missions, confirmed that this was lunar rock. The mineral compound found in this mete-orite, and also finds made in the Sahara Desert, were identical to those found in the pale-colored formations on the Moon. These 'highland' features, easily discernible when looking at the Moon, are composed of the calcium-rich mineral anorthite. The darker regions on the Moon's surface, or seas, called 'maria,' are basalt based in composition.

Of all the lunar meteorites found on Earth, it would seem that rocks don't just herald from one area of the Moon either, even including the far side of the Moon, the face constantly turned away from our gaze.

Mars

The composition of this next batch of meteorites suggests that they neither heralded from asteroids nor from the Moon, but from the Red Planet itself. Whereas all previous meteorites have an advanced age associated with them, these rocks have relatively young ages, around 1300 and 165 million years old, compared to their counterparts that date as far back as 4000 million years.

Their textures are different, igneous (formed from melted rock), implying that the body from which they came has been active more recently, in complete contrast to an asteroid or our own Moon. The ratios of the elements found within these meteorites are different from those of asteroids and those of the Moon. The elements show that it is highly likely that these meteorite finds come from a planet-sized body. All asteroids by their size are too small to remain hot for very long, whereas a larger body such as a planet can keep melting the rock for a longer period of time.

However, it was the gases trapped within these meteorites that confirmed rocks born of Mars had made it to Earth. The makeup of these Martian meteorites included black glass, formed by a shock melting, and probably produced following the impact of another body on Mars, throwing this particular chunk of rock clear of the planet's surface and far into space. The air trapped within the black glass matches the composition of the air on the surface of Mars, which was measured by the U.S. Viking lander probes when they visited the planet in 1976.

Martian meteorites have been collected from all around the world—from France, India, Egypt, America, Nigeria and Brazil—again confirming the general wide distribution globally of not just Martian meteorites but all types of meteorites.

Martian meteorites falls into one of three different groups, named after three of the best-known meteorites of this type—Shergottites, Nakhlites and Chassignites, referred to as SNC meteorites. Shergottites are the most common, accounting for by far the largest proportion of finds. Named after the Shergotty meteorite that fell to Earth at Shergotty (now Sherghati) in the Gaya district of India on August 25, 1865, there are three subgroups based on chemical content: basaltic, olivine-phyric, and lherzolitic.

The Shergotty meteorite is composed of solidified magma, produced during great volcanic activity on Mars some 4 billion years ago. Weighing 5 kg, with the meteorite's main mass a resident at the Museum of the Geological Society in Calcutta, there are similarities in composition to terrestrial dolerites. Its fall to Earth was witnessed and logged at 09:00, with an accompanying description of several detonations being heard.

Nakhlites have a very different composition to Shergottites and are thought to be considerably older, their origin probably stemming from the impact of a large body on to the surface of Mars. They are named after Nakhla, a meteorite that fell in El-Nakhla, Alexandria, Egypt, on June 28, 1911. The fall of the Nakhla meteorite, also (curiously) logged at 09:00 like the Shergotty fall, was witnessed by many of the local residents living in small settlements, describing its descent from the northwest accompanied by a trail of white smoke and loud booms. Meteorite debris was collected from a wide area, with 40 pieces collected in total. The stones ranged in size from 20 to 1813 g, with an estimated total weight of 10 kg.

The Nakhla meteorite has been something of a headline maker. Apart from the fall itself, a legend has sprung forth about the Nakhla dog, which, according to a local farmer, was struck by a fragment of the rock and reduced to ashes. There remains doubt over the story and its credibility, but it is, nonetheless, a great talking point.

More seriously, the Nakhla meteorite became the center of much controversy. In March 1999, examination of the meteorite using a scanning electron microscope (SEM), by a team from NASA's Johnson Space Center, revealed small rounded particles within the rock. The researchers, led by Dr. David McKay (1936–2013), suggested that the particles were in fact mineralized remnants of bacteria that once lived on Mars. Various amino acids were also discovered, but there was disagreement over their origin.

In February 2006, further research conducted on a sample of the meteorite held by London's Natural History Museum found a carbon-rich substance filling cracks and channels within the rock. There was a remarkable resemblance between the substance found and the effects of bacteria observed in Earth rocks.

Chassignites meteorites have similar formation ages to those of nakhlites. Named after the Chassigny meteorite that fell on Chassigny, Haute Marne, France, on October 3, 1815, at 08:00, one of the most striking composition differences is the presence of noble gas, probably indicating its source at the mantle of a larger body. The meteorite was observed falling, accompanied by a cloud of gray smoke and a succession of loud sonic booms. One fragment fell just 400 m from a local resident who had witnessed the meteorite's descent. A small impact crater had formed with freshly displaced soil and, all around it, fragments of the rock that proved hot to the touch. During extensive searches conducted across the area in which it fell only 4 kg in total weight was recovered.

Hoba Meteorite

The record for the largest meteorite to have been discovered on Earth is held by the Hoba meteorite, discovered on the farm Hoba West, near Grootfontein in northern Namibia, in 1920. Whilset plowing a field with an ox, the owner of the farm, Jacobus Hermanus Brits, had come to a dead stop amidst the noise of scraping metal beneath the plow. Having scratched away at the surface, he discovered the meteorite underneath.

Weighing 60 tons and resting where it fell, the meteorite is estimated to have fallen some 80,000 years ago. An ataxite meteorite, its original weight is thought to have been more in the region of 66 tons, but vandalism, trinket hunters, and weathering have taken their toll. Despite this, the meteorite now lies at the center of an amphitheater of concrete benches around it, and whereas in most cases rocks of this nature go to museums, in this instance the museum came to Hoba.

The Hoba meteorite is quite extraordinary in shape, more or less a cuboid, slab-like, with estimates estimating the stone to being 200–400 million years old. Apart from its curious shape, its size upon entry into Earth's atmosphere would make for a formidable force, and yet there is no impact crater. Theories abound to account for this, including one that it literally bounced along the surface of Earth, like someone skimming a stone on a pond. Declared a national monument in 1955, perhaps the Hoba meteorite entered the atmosphere at an incredibly acute angle, with only its sheer mass preventing it from ricocheting off into space.

5. Fireballs

Defining Fireballs

The official definition of a fireball, according to the International Astronomical Union (IAU), is that of "a meteor brighter than any of the planets." Although giving the observer a fair gauge of the increased brightness from a meteor to a fireball, it does leave out much when considering some of the elements of a sighting.

The International Meteor Organization (IMO) defines a fireball as having recorded −3 or above in magnitude, or brighter if seen at the zenith. The IMO, an amateur organization that studies meteors, has made allowances for the position and sighting of the meteor, correcting the visual discrepancy for brightness. It does so by taking into account the distance between the observer and sighted meteor if the meteor is, for example, spotted on or near the horizon compared to near-zenith.

For example, a meteor sighted at 5° above the horizon, with a magnitude of −1, would be classed as a fireball. The reason for this is that if the same meteor were directly overhead when sighted, it would register a magnitude of −6, enough to move its classification from a meteor to a fireball.

If the brightness of the meteor exceeds −14, it is reclassified as a bolide (derived from the Greek *bolis*, meaning a "missile" or "to flash"). This is an extremely bright meteor, possibly paralleling our own Moon for twice our lunar companion's reflected brightness.

Some fireballs have been known to produce an audible noise on entry into Earth's atmosphere, with a proportion of fireballs disintegrating into fragments. There is also a larger size of fireballs, known as a fireball meteoroid. These larger bodies are more likely to survive the passage through the atmosphere, striking Earth's surface.

A fireball meteoroid presents a very different type of fireball, where size rather than brightness matters. However, it is 'size' once again that triggers debate.

© Springer International Publishing AG 2017
J. Powell, *Cosmic Debris*, Astronomers' Universe,
DOI 10.1007/978-3-319-51016-3_5

In 1962, the IAU defined a meteoroid as "a solid object moving in interplanetary space, of a size considerably smaller than an asteroid, and considerably larger than an atom". This definition has been challenged, notably by Martin Beech and Duncan Steel, who, writing in the *Quarterly Journal of the Royal Astronomical Society*, proposed a meteoroid be classed as a body between 100 mcm and 10 m across.

However, with the discovery of asteroids measuring below 10 m in size (based on the criteria an asteroid much reach), Beech's and Steel's definition was challenged by Alan Rubin and Jeffrey Grossman, who moved to change to the measurements to 10 mcm and 1 m in diameter, in order that some distinction could be made. Rubin and Grossman, writing in the *Meteorites and Planetary Science Journal*, argued that the minimum size of an asteroid is given by what can be discovered from Earth-based telescopes, so the contrast between meteoroid and asteroid is fuzzy.

Objects smaller than meteoroids are classified as either micrometeoroids or cosmic dust. Around 15,000 tons of meteoroids, micrometeoroids, and different forms of cosmic dust enter Earth's atmosphere every year, a vast percentage of which burn up harmlessly in the atmosphere.

Light from fireballs is generated by the heating of the body as it enters Earth's atmosphere. The friction caused by the movement of the body through the atmosphere heats its surface to a temperature of 2726 °C, liquefying the object. A region of the atmosphere surrounding the body also becomes ionized, and the gas begins to glow. The trail generated from the now fireball status of the body normally starts out as a shade of white, fading to a red as the speed of the fireball becomes slower, and the ionizing effect is subsequently reduced. The colors generated give us an indication of the composition of the body, nickel producing a green color, sodium a yellow, and magnesium a blue-white appearance. The red is generated from the atmospheric ionization of nitrogen and oxygen, plus other gases.

Depending on the angle of entry, fireballs can shed weight at alarming rates. At 19 km/s, and given a 45° entry angle through the atmosphere, 80% of the fireball's original mass can survive the passage potentially intact, although the stresses placed upon it may cause it to disintegrate and ultimately fragment before striking the ground. Increase the speed of entry to 37 km/s and only 50% remains of the original body.

It is generally considered that objects weighing less than 10 metric tons lose their original velocity at an altitude of 16 km, literally being aero-braked by the atmosphere. For the rest of the descent, the body goes into freefall at a rate of 32 ft/s. As it speeds downwards, it is here, in the majority of cases, that whatever is left of the mass disintegrates into a shower of debris. Should a sizable chunk survive, impact would occur at surface level at a speed of 8368 km/h, making for a substantial crater.

Crackle, Swish, and Hiss

Apart from the bright, often colorful display, there have been many recorded incidences of a fireball making a noise to accompany its flight trajectory. However, the association of noise is not confined to fireballs, with larger meteors and meteorites across the board making similar audio disturbances—a sonic boom typically arriving seconds after the light is seen.

Reports of "crackling," "swishing," and "hissing" have all been documented, although the actual noises themselves could be the same noise, simply interpreted differently by different observers. One observer's crackle could sound like a swish to another.

These audible outbursts have also been witnessed with intense displays of the Aurora Borealis, with "crackling" being used quite frequently to describe the sound. Scientists at NASA have proposed that the turbulent wake generated by the likes of meteorites and fireballs interact with Earth's magnetic field, in turn generating pulses of radio waves. As the trail left by the entry of the body dissipates, megawatts of electromagnetic power are released, with a peak in the power spectrum at audio frequencies, thus making that part of the body's entry into the atmosphere audible as well as visual. However, this theory is not well supported, with those who contest it proposing that the sounds we hear are from the body itself, and not from its interaction with its immediate surroundings.

Researchers at Aalto University in Finland claim to have located where the sounds associated with the Aurora Borealis, or northern lights, are created. Using three separate microphones at the focus of parabolic dish antennae in an observation site where the aurora sounds were recorded, they identified noises similar to crackles and muffled bangs, 70 m above ground level, which only lasted a short period of time. The simultaneous measurements of the geomagnetic disturbances at the observation site made by the Finnish Meteorological Institute showed a typical pattern generated during occurrences of the northern lights.

Professor Unto K. Laine, an expert on speech technology from Aalto University, stated "our research proved that, during the occurrence of the northern lights, people can hear natural auroral sounds related to what they see," adding "the source of the sounds that are associated with the aurora borealis we see ... are likely caused by the same energetic particles from the Sun that create the northern lights far away in the sky. The geomagnetic disturbance produced seems to create sound much closer to the ground." However, as to the details of the sounds themselves, that remains a topic for debate, with researchers believing there are several mechanisms behind their creation.

One theory that could explain the noises, but is yet to be widely accepted, is that of what is called brush discharge. According to this idea, the ionization effects that produce the aurora are technically reaching that of ground level, but the intensity at low altitudes is not strong enough to produce any visible display to the observer. This intensity at lower altitudes causes a buildup of static electricity on nearby objects, which intermittently discharge into the atmosphere. In effect, microscopic bolts of lightning are created, which would account for the type of atmospheric sounds being heard. The effect would be stronger on long, thin, dry objects, which are best at bleeding off excess charge. This brush discharge theory has its merits, but it is not universally accepted.

Perhaps a plausible explanation for the sounds, not just associated with aurora but with meteors and meteorites, is electrophonic transduction. Certain low frequency radio waves have the same frequency as sound waves, with the aforementioned long, thin objects (grass, long hair) acting as conductors, like antennae, for these radio waves. VLF radio waves have been found to be produced by not only aurora but also by meteors.

Chelyabinsk Meteor

We move now from the −14 magnitude of a bolide to a super-bolide, reaching an apparent magnitude of −17, a feat achieved by the much documented Chelyabinsk meteor (Fig. 5.1).

The Chelyabinsk meteor was the result of a near-Earth object (NEO) , in this case a large asteroid, entering Earth's atmosphere over Russia on February 15, 2013. The meteor went undetected until it was actually in Earth's atmosphere, which by its nature is alarming by itself, although the entry point was close to the Sun, which may have gone some way to explaining why the body was not picked up sooner.

Traveling at an estimated speed of nearly 43,000 mph (69,000 kmph), this super-bolide blazed through Earth's atmosphere, exploding in an air burst over Chelyabinsk in the southern Ural region of Russia. The super-bolide generated a bright flash of light that was brighter than the Sun, even to observers 48 km away from the point of entry into Earth's atmosphere.

The angle of the object's entry, while causing significant alarm, personal injury and property damage, is perhaps the saving grace from what could have been a much more serious incident. Exploding at a height of 97,400 ft, and weighing an estimated 10,000 metric tons, the object must have made for a terrifying sight. Another sobering thought is to imagine what would have happened if the Chelyabinsk meteor had remained intact at an estimated 55 m across, its full mass striking Earth's surface!

Fig. 5.1 This photograph of the meteor was taken by a local M. Ahmetvaleev as it streaked across the sky, Chelyabinsk, Russia, February 15, 2013. The small asteroid was estimated at 55 to 56 feet wide, (17 to 20 meters). (Image copyright M. Ahmetvaleev. Courtesy of NASA)

The Chelyabinsk meteor produced a huge shockwave, with most of the energy (the equivalent of which has been estimated to be 30 times that of the atomic bomb detonated in Hiroshima) absorbed by the atmosphere. However, what energy wasn't absorbed by the atmosphere resulted in the shattering of glass in an estimated 7200 buildings, spanning six cities, resulting in approximately 1000 people requiring medical attention. According to NASA scientists, the shockwave was so powerful that it traveled twice around the world.

Although cuts and bruises were commonly reported following the impact, other more bizarre incidents were also recorded. An Internet-based survey conducted shortly after the event by Dr. Peter Jenniskens, a Dutch and American astronomer working at the Search for Extra-terrestrial Intelligence Institute (SETI), in Mountain View, California, revealed some interesting results.

Along with a team from the Russian Academy of Sciences who attempted to simulate at ground level the path of the Chelyabinsk meteor's descent through the atmosphere, the Internet survey of people who were in some way medically affected by the event found that, out of 374 that responded, the most common complaint related to the eyes. Of the respondents, 180 said that their eyes hurt, with 70 claiming temporary blindness from the initial explosion of light. However, more intriguingly, 20 reported sunburn—so bright was the flash from the meteor. One even claimed that his skin had actually peeled. To counter-balance this claim, Jenniskens, a meteor expert, added: "But he [the respondent who claimed his skin peeled] was also in a snowed-in landscape, and snow is very efficient at scattering UV light. That may have helped."

The Chelyabinsk meteor is the largest known natural object to have entered Earth's atmosphere since the Tunguska event of 1908, which destroyed a remote forested area of Siberia. Since this reality check of 2013, scientists have strived to find the origin of the Chelyabinsk meteor, details of which have been slowly revealed over the years since that fateful day.

Discovered on March 11, 2011, and part of the Apollo NEO group of asteroids, 2011 EO40 seems to be the likeliest candidate as the parent body from which the Chelyabinsk meteor was born. It is thought that a chunk of the 2011 EO40 asteroid, which measures 656 ft wide, is responsible for the incident, the large fragment possibly gouged out of the asteroid as result of a collision with another body. 2011 EO40 regularly crosses Earth's orbital path and is listed as a potentially dangerous asteroid or potentially dangerous object (PDO). Considering just a fragment of the asteroid struck Earth in the form of the Chelyabinsk meteor, the concern attached to the body is well-founded.

There is reason to believe that perhaps it was not a collision that produced the Chelyabinsk meteor, with the gravitational influences from both the planets and our Sun the possible cause, literally ripping a sizable chunk away from 2011 EO40. If this is true, then other fragments could well break free in the future. Tracking known NEOs is one thing, but, as with the undetected entry of the Chelyabinsk meteor of 2013, tracking breakaway debris from a parent body is another.

The link between 2011 EO40 and the Chelyabinsk meteor was made by two astronomers working at the Complutense University of Madrid. Professor Carlos de la Fuente Marcus, along with his brother Raul, discovered statistical evidence to support the possibility of a Chelyabinsk cluster, closely associated not only with the gravitational field of the Earth-Moon system but affected also by Venus and Mars and by Ceres, the largest occupant of the Asteroid Belt, considered by its size to be a dwarf planet. Their research estimated the cluster to be in the region of 20,000—40,000 years old, with 2011 EO40 the most likely candidate for the Chelyabinsk meteor event, although purely by coincidence, a 164 ft-wide asteroid, 2012 DA14, actually passed Earth hours before 2011 EO40 (Fig. 5.2).

Professor Carlos de la Fuente Marcus commented that originally there were 20 candidates in the so-called Chelyabinsk asteroid family that could have been responsible for the event, with more pieces expected to break free in the future. However, a second incident relating to the same body in this case was unlikely.

Although the Chelyabinsk meteor surprised the world, the actual reconstruction of its path into Earth's atmosphere didn't require much effort, as the event occurred at a time when incidents of this nature were often being caught on camera. Be it the camera

Fig. 5.2 The far side of our defenseless Moon, bereft of atmosphere, and offering no resistance to space debris. Picture taken by MoonKAM, (Moon Knowledge Acquired by Middle school students), onboard the Ebb Spacecraft (courtesy of NASA)

on a mobile telephone, or dash-mounted car cameras, never before has the sky been so inadvertently monitored by so many.

A 570 kg rock, believed to one of the largest pieces of the Chelyabinsk meteor, was put on show to the general public after the event. Exhibited in the Chelyabinsk History Museum in the Southern Urals, the splinter of rock was recovered from the bottom of Lake Chebarkul. Shown live on Russian television, a team pulled this 1.5 m-long rock from the icy depths, having been encased while underwater with protective wrapping. Weighing in at 570 kg, the rock not only broke the scales when it reached that weight but also split into three pieces.

Dust to Dust

The explosion of the Chelyabinsk meteor upon entry into Earth's stratosphere also caused the depositing of hundreds of tons of dust at high altitude. These deposits were closely analyzed by the NASA-NOAA Suomi National Polar-orbiting Partnership satellite, the findings revealing that the material left by the explosion had formed a thin but cohesive and persistent stratospheric belt of dust around the globe.

Using data gathered from the Ozone Mapping Profiling Suite instrument's limb profiler on the satellite, atmospheric physicist Nick Gorkavyi, along with colleagues at NASA's Goddard Space Flight Center, combined the findings with computer modeling to simulate how a plume from a bolide explosion would evolve as the stratospheric jet stream carried it around the northern hemisphere.

Less than four hours after the explosion, the satellite detected the plume at an altitude of 40 km. The plume moved quickly eastwards at speeds of around 306 km/h. A day later, the satellite continued to track the plume as it traveled eastwards, reaching the Aleutian Islands. Although smaller particles within the debris were able to stay aloft within the plume, it was noted that heavier particles were losing altitude and speed. Four days into the explosion, the higher, faster altitude particles had completed a circuit of the globe, arriving back at the point of entry at Chelyabinsk. A detectable belt of bolide particles remained in evidence for at least three months after the event, with the computer model simulations and then subsequent actualities of the satellite findings 90 days on showing complete agreement in the evolution of the plume's movements and structure throughout its trek around Earth.

Not only had the findings achieved the accolade of being the first space-based observations of the long-term evolution of a bolide plume, through the work of computer modeling it had given scientists an insight into how possible larger explosions may perform. The Chelyabinsk meteor was considerably smaller than the body that saw the demise of the dinosaurs, but the principal applied for plotting the plume and subsequent distribution of debris provided yet another useful tool in analyzing the aftermath of a strike on the environment.

Every day, small debris from various sources encounter Earth, becoming suspended high in the atmosphere. Despite the debris, the environment remains relatively clean at this level when compared to a stratospheric layer below, where the environment is far from clean, harboring abundant natural aerosols from volcanoes and other larger and less sparse particles.

The Chelyabinsk meteorite is a firm reminder of what can happen should a sizable chunk of debris enter Earth's atmosphere. A firm reminder certainly, but also perhaps thought provoking enough to consider those that have narrowly missed us.

6. Near-Earth Objects

The Recognized Threat

The chances of a random piece of rock striking Earth still exists, but probably to a much reduced level of threat. During the early years in the creation of our Solar System, there was plenty of debris that had yet to settle into some form and order, with constant collisions of large bodies before any structure was imposed. As time passed and natural order and stability were imposed, collisions were reduced in number, and, with it, the chances of being hit subsequently fell. The remnants of this settling down phase were either ejected into deep space or conformed to some kind of symmetry, even though within that symmetry the threat of being hit remains.

One such chaos within the order is presented to us by near-Earth objects. NEOs have been in existence since the formation of the Solar System, and, while knowledge has been gathered as to their size and orbital paths, it is unlikely that a point will ever be reached when the Solar System settles down into a steady and rhythmic accord, where everything within it behaves in a pattern that is constant, predictable, and hazard-free. There is a state of flux within stability itself, and that's without considering influence from outside of the Solar System. We pride ourselves in the predicting and modeling of astronomical occurrences, from our Moon's phase to the occultation of a star by a planet, but there remain many instances where the application of modeling simply can't be applied with a healthy degree of certainly that what we are told matches that which eventually transpires.

The universe cannot function in a healthy manner on such a diet of predictability. A random element must be introduced so that the so-called existing order is allowed to grow and evolve. This applies to NEOs, for here, monitoring affords us relative peace of mind that some governance on order is maintained. However, it does not account for a random collision in the Asteroid Belt, or a

J. Powell, *Cosmic Debris*, Astronomers' Universe,
DOI 10.1007/978-3-319-51016-3_6

comet release from the Oort Cloud. It does not account for the movement of said bodies and their eventual resting place. We can observe, but we can do little more.

'Near-Earth objects' is the collective term for all objects in the vicinity of Earth, with near-Earth asteroid (NEA) defining that particular object clearly and distinctly as an asteroid, rather than any other body, such as a comet. By the end of August 2016, 14,686 NEOs had been logged and their positions recorded, with 1729 of these classed as potentially hazardous asteroids. If discussing the potentially hazardous phenomenon in its entirety, the term used would be potentially hazardous objects (PHOs). However, careful study is required to establish the exact classification of the body (Fig. 6.1).

With regard to the NEAs in particular, much research has been undertaken in an attempt to understand what makes them 'tick,' not only in terms of their chemical composition, about which spacecraft have enlightened us enormously, but, most importantly, their movements over a period of time. One such research project has been investigating spin states of NEAs,

Fig. 6.1 When first observed by NASA's Neowise Team on December 31, 2013, Comet Catalina gave the distinct appearance by its movement of being that of an asteroid rather than a comet (courtesy of NASA)

analyzing the fundamental core properties of designated asteroids. Understanding the pole directions and spin rates of this sample could be key to our overall knowledge as to how the population works as a whole. Spin rates, with regard to shape and size, can be helpful in predicting the long-term trajectories of such objects, not only providing crucial data as to where they are expected to end up but also passages close to Earth in the near future, which will offer us an opportunity to send probes to investigate and analyze such bodies close up. Also, when the technology exists, we could possibly redirect or steer them clear.

The ability to influence the direction of a body is both a comfort and a worry. Knowing that in the future there would be some possibility of halting or deflecting an incoming rock is a great reassurance, although a body of a certain size would be a challenge. The worry element is the disruption, albeit for Earth's benefit, to the order of the Solar System. Altering the path or direction of one rock could create an imbalance, as if the order of the Solar System had been tampered with.

The NEAR Mission

Whereas lunar and planetary exploration were the primary targets for exploration over many decades and by many countries, the focus has now switched into understanding other components of the Solar System, with particular attention paid to the study of NEOs. Launched on February 17, 1996, the Near-Earth Asteroid Rendezvous (NEAR) spacecraft made successful flybys of asteroid 253 Mathilde on June 27, 1997, and asteroid 433 Eros on December 23, 1998 (Fig. 6.2).

Discovered by Austrian astronomer Johann Palisa on November 12, 1885, this main-belt asteroid, measuring around 50 km across, travels in an elliptical orbit and takes around four years to orbit the Sun. A portion of 253 Mathilde's orbit sees the asteroid reach the limb of the belt, but still well within the parameters of both Mars and Jupiter, and not intersecting either planet's orbit.

The rate of rotation for 253 Mathilde exceeds 17 days, making the body a relatively slow rotator, with other asteroids normally

Fig. 6.2 Artist's rendering of the NEAR (Near Earth Asteroid Rendezvous) spacecraft encounter with the asteroid Eros (courtesy of NASA)

completing a rotation between 2 and 24 h. Only two other aster-oids, 1220 Clocus and 288 Glauke, are known to have longer rotational periods. This slow rotation was at first a puzzle, but a credible explanation would stem from the fact that the asteroid has a companion, its own satellite.

NEAR encountered 253 Mathilde at a distance of 1200 km but, despite taking over 500 images, was only able to capture one hemisphere. The very dark appearance of the asteroid is compa-rable to fresh asphalt, hence the low albedo, reflecting just 4% of its surface. This C-type asteroid's composition, essentially domi-nated by phyllosilicate minerals, is shared by carbonaceous chon-drite meteorites. The NEAR spacecraft was also able to establish that 353 Mathilde consisted of essentially a collection of very loosely packed debris, meaning that up to 50% of the asteroid's inner volume is space.

NEAR's photographs revealed 254 Mathilde to be a crater-strewn world, with two large craters, Ishikari, 30 km across, and Karoo, 31 km across. The appearance of the craters would seem to imply that some of the asteroid's mass had been lost in collisions, resulting in these defined gouges. The albedo deep within all of the photographed craters mirrored that of the surface, with no discernible shift in brightness levels even at these lower levels. At a distance of 17,895 km, NEAR was able to reveal one particular impact crater that was nearly 10 km wide.

The primary mission of NEAR was to encounter Eros late in 1998. The rendezvous with 253 Mathilde was an afterthought from scientists on Earth, only discovering a year before NEAR's launch that it was possible not only to visit Eros but to rendezvous with 253 Mathilde. The deliberations over targeting another asteroid and not just Eros could have jeopardized the whole mission. Should they take a risk in using precious fuel to divert NEAR to 253 Mathilde then go on to Eros? The great uncertainty as to whether NEAR could survive the first asteroid encounter would have to be considered, given the close contact with all manner of debris associated with close flybys. However, the gamble was taken, and it duly paid off, with this rather low-cost mission of NASA's delivering invaluable scientific and photographic data of not just one but two asteroids.

The slow rotation of an asteroid merits further investigation. With 253 Mathilde taking 17 days to rotate, perhaps examination of other slow-rotating asteroids would also show a companion? If so, would the presence of a satellite or indeed a possible binary partner account for all asteroids that rotate slowly? Could any combination have a crucial bearing on orbital deviations over a long period of time, perhaps steering them on a collision course with a larger body, such as a planet?

The Eos Family

The first of the aforementioned asteroids, 1220 Clocus, is a member of a group of asteroids known as the Eos family, with photoelectric studies conclusively showing two noticeable periods where there is variation in reflective light output. Could 1220 Clocus have a satellite, or could it be a binary asteroid? That would explain the noticeable spikes in light. Or is it just that, at certain times, the side presenting itself towards Earth has significantly less reflective material than the rest of the body? This proved to be an unlikely theory, as no such matter difference on an asteroid's surface could account for that.

The Eos family, or Eaon family, is a prominent group of asteroids that orbits within the Asteroid Belt between Mars and Jupiter. Named after one of its members, 221 Eos, the bulk of the family hold a steady and consistent orbit around the Sun, although a number of bodies within the family have rebelled and left the group. These more radical elements are noticeably younger within what is essentially an old collection of fragments, the result of an ancient collision. It is possible that these younger members were perhaps captured or realigned into the group by planetary resonance, with their orbit less stable and less cohesive than more established family debris, possibly orbiting in a path wider and more susceptible to influence. In turn, their wider orbital path would leave them exposed to outside influence, allowing a number of the fragments to break away. As a further development, those that haven't broken away from the fold appear to have informally clustered in their own groups.

Of the 4,400 members of the Eos family, the majority fall into the S-type category of asteroid, although examination of the group members using infrared showed differences within the category itself. As a result, and in order to keep some reign on the classification, the family was given its very own category: K-type asteroids. C-types are the most popular category of asteroid composition, S-types are second, and approximately 17% of all asteroids fit into the K-type category.

Under the Small Main-Belt Asteroid Spectroscopic Survey (SMASS) classification, K-type asteroids are further grouped together with other similar strand members, the entire range being placed under the S-type. The SMASS was a project initiated at the Massachusetts Institute of Technology (MIT) in 1990 with the purpose of studying small asteroids within the main Asteroid Belt, in order of their spectral shape and color. This, together with their albedo rating, defines what types of asteroid they are.

Before SMASS, the first taxonomy used on asteroids was proposed by American astronomer David J. Tholen in 1984. Named the Tholen Spectral Classification, it used a combination of data collected during the Eight-Color Asteroid Survey (ECAS) in the 1980s, plus albedo measurements. Tholen based his classification on 978 asteroids, with 14 different asteroid types established. In turn, SMASS analyzed 1447 asteroids, classifying the findings into 26 different asteroid types.

In 1918, Japanese astronomer Kiyotsugu Hirayama (1874–1943), while studying at Yale University, discovered that more asteroid orbits were similar to one another than chance would allow. Hirayama concluded that therefore families of asteroids existed, with 19 members of the Eos family. By 1993, this figure had risen to with 289.

Hirayama's studies were fundamental in establishing that certain groupings of this kind exist, with the orbits of 790 asteroids studied. By 1919, five Hirayama families, as they were collectively known, were established. Apart from the Eos, which at this point had climbed from 19 to 38, there was the Themis family with 31; the Koronis family with 23; the Maria family with eight; and the Flora family with 81.

Kiyotsugu also hypothesized that these families, of which many more have subsequently been discovered, were the result of catastrophic collisions with a parent body, an interpretation still widely accepted in the astronomical community today. In the early formation of the Solar System, significantly larger bodies most likely existed where only fragments are present today, with many years of further collisions taking place before an historical chain presented itself. The chain consisted of larger bodies that survived the encounters intact, right down to the tiniest fragment that perhaps still travels in the orbit of the Asteroid Belt. The collision that resulted in the formation of the Eos family places its members' collective age at 1.1 billion years old.

Discovered by Karl Theodor Robert Luther (1822–1900) in 1890, and named after Glauke, a daughter of Creon, who was king of Corinth in Greek mythology, 288 Glauke has a rotational period of 50 days. 288 Glauke is an S-type asteroid, and although comparisons with other S-type asteroids place it in a class of its own for rotation periods, there is a similarly slow but not as slow same-class asteroid, 4179 Toutatis, with a rotation period of just over seven days.

4179 Toutatis

First sighted on February 10, 1934, and named as object 1934 CT, 4179 Toutatis was considered lost as an asteroid for many years. It was presumed that the asteroid was destroyed in a collision with another body or, if not destroyed, had its orbit radically altered so that it left the Solar System, never to return. A fair proportion of asteroids and other such debris are only ever recorded once then lost forever. However, on January 4, 1989, French astronomer Christian Pollas rediscovered it, giving it its current name Toutatis, after the Celtic god of tribal protection.

Pollas, who also has a main-belt asteroid named in his honor, has many a find to his name, including 4179 Toutatis, which he co-discovered with Belgian astronomer Eric Walter Elst. Elst's find count is staggering, literally thousands, including 4486 Mithra, which, like 4179 Toutatis, is a potentially hazardous object. Furthermore, 4179 Toutatis is both an Apollo and an Alinda, with its orbit intersecting that of Mars. Earth and Jupiter have affected the asteroid's orbit, resulting in frequent approaches to Earth.

The asteroids of the Alinda family are held in their orbit by a 1:3 resonance with Jupiter, which results in their being close to a 4:1 resonance with Earth. Any asteroid, or any object for that matter, in this resonance has its orbital eccentricity steadily increased over time by the gravitational interactions with Jupiter, until it eventually has a close encounter with an inner planet, such as Venus or Earth, subsequently disturbing this pattern and altering its path.

The first of the Apollo family to be discovered was 1862 Apollo, found by German astronomer Karl Reinmuth (1892–1979) on April 24, 1932. Apollo 1862 is a PHA itself and, most notably, the Chelyabinsk meteor was a member of this family.

In recent times, 4179 Toutatis has passed close to Earth in 2012 and 2016, with 2069 marked as the next closest approach, based on there being no future alterations to its current orbit. In 2012, the asteroid passed Earth at a distance of 6.9 million km, close enough to be spotted on Earth with high powered binoculars. Its next scheduled closest approach is November 5, 2069, when it will pass 2.9 million km away.

The rather odd shape of 4179 Toutatis would seem to indicate the merging of two distinct bodies, possibly a former binary, or the long-term fusing between two bodies of similar mass over time. The two lobes the asteroid presents (perhaps akin to a lopsided peanut), coupled with a 'tumbling' motion similar to that of 288 Glauke, would give a close-up observer a unique sight during its obscure rotational phase. As its rotation combines two separate periodic motions into a non-periodic result, the observer would see the Sun rise and set at random locations. There could be no such thing as a proper 'day' on 4179 Toutatis, its rotation providing the observer with periods of between 5.4 and 7.3 Earth days.

This tumbling motion of 4179 Toutatis is a result of the Yarkovsky or Yarkovsky-O'Keefe-Radzievskii-Paddack (YORP) effect, which describes a force that acts upon the orbital motion of asteroids and other smaller bodies whose diameter tends to be less than 64 km. It is caused by light from the Sun, which heats up the asteroid to a point where the body is capable of radiating the energy away, which in turn creates a thrust-like motion. Over a long period of time this can have an effect on the asteroid's orbit. Apart from making its presence known to Earth, 4179 Toutatis, because of its low inclination of orbit, makes frequent orbital transits, where the inner planets of Mercury, Venus, Earth, and Mars can appear to cross the Sun from the perspective of an observer on the asteroid.

Most PHAs herald from the Apollo family and, to a lesser extent, Aten asteroids. As with Apollo, the first to be discovered in this group was 2062 Aten—discovered by American astronomer Eleanor F. Helin (1932–2009) on January 7, 1976. As former principal investigator of the Near Earth Asteroid Tracking (NEAT) program, Elin recognized 2062 Aten to be a part of a group just like the other families and, although collectively smaller in number than the Apollos, still harboring a threat.

Bennu

The PHA 101955 Bennu is another member of the Apollo family, and attracts great interest because of its future Earth encounters. Discovered on September 11, 1999, by the MIT Lincoln

Laboratory-funded Lincoln Near-Earth Asteroid Research (LINEAR) program, this particular asteroid is due to pass between Earth and the Moon in 2135. In doing so, calculations show that this will alter the path of Bennu, potentially putting the asteroid on a collision course with Earth during a later encounter. Measuring just under 500 m and traveling at around 63,000 mph (101,000 km/h), a collision of this magnitude is likely to cause considerable damage—not on the level of an extinction event but the equivalent of a nuclear warhead exploding, or that of a magnitude 7 earthquake (Fig. 6.3).

Passing every six years, Bennu is rated number three on the Palermo Technical Hazard Impact Scale, a scale defined by using the probability of impact and the kinetic energy released should it strike. Bennu and others like it are also under constant scrutiny from an automatic Near-Earth Asteroid Collision Monitoring System (SENTRY), continually watching their movements for the possibility of orbital changes and subsequent future impact.

Fig. 6.3 The results of past impacts present us with the pitted, scarred, and cratered lunar surface of our closest companion. Picture taken by Apollo 8, December 21–28, 1968 (courtesy of NASA)

Earth's Eyes on the Skies

Understanding how the field of space debris works can provide insight into dealing with the escalating problem that these fragments pose. However, even with this greater understanding, the task of monitoring and tracking debris is a substantial undertaking, particularly so given the rate of new finds and the concern over the proportion of objects that are not yet cataloged. One of the tools at our disposal is CoLiTec (CLT), a software package dedicated for the automatic discovery of any moving objects such as asteroids, comets, and other cosmic debris. Data processing for Comet ISON was carried out by CoLiTec following its discovery on 21 September, 2012 by Vitaly Nevsky and Artyom Novichonok (Fig. 6.4).

Orbital models have been developed, the most intricate of which is NASA's EVOLVE. Monitoring the predicted growth of low-orbit space debris, their mass, trajectory, and velocity, EVOLVE and similar programs project a future risk assessment.

Fig. 6.4 Comet ISON shines brightly in this image taken on the morning of November 19, 2013. This is a 10-second exposure taken with the Marshall Space Flight Center 20″ telescope in New Mexico. Image credit: NASA/MSFC/MEO/Cameron McCarty (courtesy of NASA)

Several NASA-run surveys sweep 6000 square degrees of sky every night looking for NEOs. Another NASA project, using the Wide-field Infrared Explorer (WISE), has been scouring the skies since 2013.

The Inter-Agency Space Debris Coordination Committee (IADC) has also been set up to evaluate future risk, with many other countries now taking steps to develop policies with regard to space debris as, after all, it is a global issue. With the use of hindsight, the problem of space debris is also not just confined to NEOs, with human-generated debris an issue that should have been considered since the launch of the first satellite into space. This apparent neglect has seen a staggering amount of debris left for future generations to have to address and, as the decades pass, the litter that orbits our Earth continues to grow at an alarming rate.

Despite the show of concern about debris, the bombardment continues on a daily basis, not just from oddments of our own debris but from rocks large enough to make it through the atmosphere, many of which fall unseen on Earth's vast oceans or on wasteland and Arctic tundra. The Minor Planet Center in Cambridge, Massachusetts, conducts one of the oldest ongoing surveys to date, having recorded and cataloged the orbits of many bodies since 1947. The attention to NEOs and acknowledgement of those bodies that cause a threat closer to home spawned NASA's Near Earth Object Program, part of the Spaceguard program. Spaceguard is a collection of affiliated programs working in unison with regards to NEOs, their whereabouts and subsequent tracking. The UK has its own Spaceguard, the National Near-Earth Objects Information Centre, based in Powys, Wales.

Another NASA project, in conjunction with the United States Air Force, is LINEAR. The project, based at the White Sands Missile Range in Socorro, New Mexico, is made up of two 1 m telescopes and one 5 m telescope, and collectively they have been responsible for discovering thousands of objects a year. LINEAR uses very specialized equipment known as Ground-based Electro-Optical Deep Space Surveillance (GEODSS) telescopes.

In 1984, the Spacewatch project became the first to detect and discover asteroids and comets with CCDs, as opposed to photographic plates or film. A 0.9 m telescope and a 1.8 m telescope were used, the latter on fainter objects that become fainter after their discovery. Based at the Kitt Peak Observatory in Arizona, Spacewatch scours the skies for NEOs on a constant basis.

Drawing on international expertise is the German space agency's (DLR) Institute of Planetary Research in Berlin. The DLR spearheaded NEOShield, a project set up to analyze realistic options for the prevention of collisions between Earth and NEOs. The main objective for consideration is whether to push or pull an approaching object, with such ideas including a "gravity tractor" to pull the approaching object away from Earth. Another idea on the table, which is far from new, is "blast deflection"—the detonation of a possible nuclear device near the object to literally deflect its course.

Various craft have visited NEOs, including a visit to asteroid (25,143) Itokawa. Itokawa is a 600 m potato-shaped asteroid named after Hideo Itokawa (1912–1999), a Japanese rocket pioneer. Launched on May 3, 2004, from the Kagoshima Space Center, the Japanese probe Hayabusa rendezvoused with the asteroid on September 12, 2005. Having landed on the asteroid after spending time matching Itokawa's orbit, a first attempt to take samples of the body failed, but a subsequent effort proved successful, with a capsule containing the findings returning to Earth on June 13, 2010. Itokawa, an S-type asteroid, revealed itself to have probably been made up from two components, with the particles collected from its surface—only the third in history to be collected from an extra-terrestrial body, following NASA's Apollo and the Soviet's Lunar expeditions.

However, despite all the efforts to detect, track, analyze and visit all these objects, in essence, should one come our way, there is nothing we can do, especially if like the Chelyabinsk meteor it is spotted incredibly late. Technology simply has not advanced far enough to deal with such a catastrophe were a body of destructive size to head our way. We are ultimately at the mercy of NEOs and, for that matter, any comet or wandering asteroid that happens to visit our region of space.

With all the investment in the world, and even with a monumental global effort, the percentages do not favor success. Much of what is written about deflecting or dealing with any celestial body has no substance, and with the threat as real today as ever we still do little, because little is all we can do. If setting up committees, projects and programs to look into the threat were as prolific as the actual planning and subsequent construction of a 'vehicle' to defend Earth, then we might have some assurance that something is being done, but it isn't, and therefore, as with the dinosaurs, we might ultimately endure the same fate.

7. Life Givers or Life Takers?

Seeds of Life

Theories abound about whether life came from space—whether our very being came from the stars. It is a possibility of such huge magnitude and significance that it cannot be ignored, but it remains a topic of much speculation and conjecture. As with all theories, the only guaranteed way to reassign it from theory to fact is to have solid, irrefutable proof that is watertight and unimpeachable.

If life did come from space, then it would perhaps be foolish to think we are alone, when across the universe other planets with life-sustaining atmospheres exist—other worlds evolving as ours does, and the possibility of another parallel type of life, similar to life that has evolved on Earth. However, to have exactly the same set of circumstances for life to grow on other worlds akin to Earth is surely highly unlikely if not impossible, even if the life was brought to this other world by the same means. What, then, of the proposed continual spreading of life throughout the universe from one planet to the next, a sequence of "seeding" planets?

From the Greek for "seeds everywhere," panspermia, first speculated on in the writings of Greek philosopher and scientist Anaxagoras (500–428 BC), is the hypothesis that states that the "seeds" of life exist all over the universe, propagated throughout space from location to location, one planet to the next. One such carrier of these seeds, in the form of microorganisms, could well be an asteroid, or a comet.

These microorganisms possibly contain the very "building blocks" of life, with necessary proteins to establish the simplest of life forms. Discovered on comets, and in meteorite fragments, these traces of amino acids and other organic compounds may well have formed in space, long before a descent through the atmosphere of a lifeless planet. However, as logical as it may seem, it remains a theory, as proving panspermia would require a massive

J. Powell, *Cosmic Debris*, Astronomers' Universe,
DOI 10.1007/978-3-319-51016-3_7

survey encompassing vast regions of space, and the exploration of other planets.

Several variations exist on the panspermia hypothesis:

- Life began once on Mars and spread to Earth via Martian meteorites.
- Life originated independently on both Mars and Earth. Cross-colonization may have then occurred.
- Life began once, on Earth, and was then propagated to Mars, where it may well have established itself.
- Life originated on both Mars and Earth. However, despite an exchange of debris in the form of rocks and dust between the two, no transfer of viable organisms occurred.
- Life did not originate on either Mars or Earth, but somewhere else and, by a series of different possible means, was brought to Earth.
- Life began on Earth and stayed on Earth, with no transfer of life to another planet.

The obvious alternative to the panspermia idea is that life began on Earth and stayed on Earth.

Despite support for the pansmeria idea as a whole evidence remains rather elusive, with only the discovery of the Mars meteorite in the Allan Hills in Antarctica on December 27, 1984, bringing the possibility of life arriving from Mars into the media spotlight. Despite this meteorite, named Allan Hills 84001, containing what appeared to be evidence of Martian bacteria, doubt set in as to the findings, and the whole premise for a link to life coming from Mars in meteorites like this were dismissed, until perhaps other discoveries are made.

Life from a Comet

There remains nothing more thrilling to the astronomer than the sight of a new visitor to our skies—a bright, impressive comet, coming reasonably but not fearfully close to Earth and, like several comets throughout history, sporting a grand and striking tail.

A traveler from the depths of space appearing in our own backyard is quite a treat, because throughout the vastness of space, somehow, this particular body made its way to our neck of the woods. In its fiery, steely glare, below the comet, a world of wondering minds stares back, some eager to categorize the stranger, dissect and evaluate the body. For others, it is a once in a lifetime look at what could well be a remnant from the beginning of time, evading capture from the gravitational pull of a star or planet, a true cosmic time traveler.

However, not all eyes welcome the visitor. Some see the stranger as a threat, a harbinger of doom, and a sign of woeful times ahead. But surely this comet is but harmless, just a great big snowball from the depths of space? Indeed, if it is not on a collision course, just passing through, then what on Earth could be sinister about it?

With the Earth passing through many cometary trails and debris during its existence, does whatever lurks within these fragments harbor something harmful, a virus or a plague? Or is the comet a potential carrier of life?

When Halley's Comet last put in an appearance back in 1985/86, British mathematician and astronomer Sir Fred Hoyle (1915–2001), along with colleague Chandra Wickramasinghe, also a British mathematician and astronomer, came to the fore with ideas that suggested comets carried key molecular material and, when in the vicinity, could possibly infect Earth with it. The idea went further, suggesting possible bacterial or viral material. Hoyle and Wickramasinghe worked on such theories long before the heightening of media attention during the return of Halley's Comet. Their belief that life on Earth is partially extraterrestrial is derived from many years' work, but probably the most controversial aspect of what they propose is that, apart from potentially delivering life, cometary debris that has entered our atmosphere over the centuries has also been responsible for major epidemics experienced across the globe.

Although intriguing, the theory remains a theory. But it does beg the question of what exactly is bound up within the comet's makeup, what possible material is locked away in the body's frozen casing—a casing that, when close to a star like our Sun, melts and fractures, revealing what has been stowed away.

Downfall of the Dinosaurs

Sixty-five million years ago, an event occurred that caused a dramatic and life-altering shift on Earth. What was thought to be a comet rather than an asteroid (for reasons that will later be explained) caused widespread and devastating destruction on what was then the kingdom of the dinosaurs. For here, among the roaming giants, an end would be met that would redirect all subsequent life since and change the whole appearance of this once lush and vibrant planet.

With a blinding light as the comet plowed into the atmosphere, the impact that followed was the equivalent in explosive force to that of 100-million-megaton bomb, a huge and devastating detonation, especially considering the destruction caused by 13,000 tons of the Hiroshima atomic bomb. The blast was catastrophic, wiping out 60% of all creatures in existence at that time, including as much as 75% of marine life. The aftermath of the impact was to cause long-term issues, with the animals not killed instantly on the initial strike left to suffer a long and protracted death, as debris from the impact was thrown up into the atmosphere, shrouding the Sun's light, causing vegetation and plant life to wither and die. In time, the globe was to become a frozen wasteland. Those animals that could survive the shift in environment fled underground. Many were already skilled at evading predators by escaping into the ground but, this time, a long passage of time transpired before the survivors ventured forth above ground again.

Fossil dating points to this incident as not being a lone or rogue occurrence, but just one of perhaps as many as nine or more episodes that Earth has suffered in its turbulent history, all of these events taking place in the last half billion years. Does this therefore mean that it is only a matter of time before another collision of extinction level occurs? Seemingly so, but not the sort of precise date and time that one could enter in a diary or on a calendar—it could happen at any point. The argument may exist that Earth's past was more volatile, and therefore open to such occurrences, as perhaps the cosmos was still in a relatively active state before any settling down occurred. In this state of flux, more debris was evident, and therefore, with a large amount of fragments flying around, was there an increased likelihood of impact? Possibly so, but dwell on this. There may well be a lot less debris around and, as science advances, the ability to spot and track it now gives us an edge, whereas before, Earth was somewhat blind. However, whether we spot it or not, we are, in truth, ultimately defenseless. All the tracking only allows us to see the face of the intruder before the act is committed, nothing more.

Why do we think that the obliteration of the dinosaurs was a comet rather than an asteroid? The fossil layer attributed to the time period, the K/T boundary (the boundary between Cretaceous and Tertiary periods in Earth's history), has evidence of the presence of iridium, a rare metal not commonly found on Earth's surface. At an early stage in the evolution of the planet, iridium sank towards Earth's core, drawn downwards by iridium's nature of sticking to iron, which in perhaps greater capacity was being sucked down in quantity at the time.

The layer of debris and particles thrown up into the atmosphere on impact contained a substantial amount of iridium, and for a thousand times more iridium to be present a massive object, probably of cometary-like proportions, must have been responsible, given the potential rate at which it was traveling and the depth of impact. Although the argument isn't all in favor of it being a comet, the presence of amino acids also in the K/T boundary would give credence to the body at least being of extraterrestrial origin, as both types were present, those more commonly associated with terrestrial life and those found in abundance in meteorites.

However, further research concluded that, because the amino acids were not discovered in the same section of the K/T boundary as one would have expected (both above and below it), it more than suggests that the amino acids were enveloped in a cometary dust trail, both in front of and after the impact. The impact and associated energy itself would have destroyed any amino acids, thus subsequently not present in the iridium layer.

Comet or asteroid, Earth's slate was wiped virtually clean, allowing for a new order to evolve. This also begs the question that had this extinction level event not happened, what exactly would exist on Earth today?

The Tunguska Event

More recently in Earth's history, one such event remains clearly marked on our timeline, that of Tunguska. For here, on the morning of June 30, 1908, in the Tunguska River Valley area of Siberia, another cometary-like object was to invade. A brilliant, flaming fireball lit up the sky in a spectacular blaze, exploding into showers of burning debris as it did so, and leveling all in its path (Fig. 7.1).

This superbolide disintegrated at high altitude, with remnants including what was left of the main projectile, and flattened an area of trees some 60 km across, with the resulting prostate wood all pointing in the direction of the impact site. In total, 80 million trees were felled like matchsticks in the Taiga forest, leveled with one almighty swipe from space. The trees were scorched rather than burned, as the initial flames from the descending object lit them while the trailing passage of wind then quickly extinguished any flames. As Earth shook, windows smashed in a town some 60 km away because of the sonic boom.

Fig. 7.1 Trees felled by the Tunguska explosion. Image credit: Leonid Kulik (courtesy of NASA)

The area of impact was thankfully sparsely populated—just a number of small settlements in what was chiefly a wooded landscape. However, one death was reported, that of a local deer herder who was thrown into the side of a tree following the blast. Another report also mentions an old man dying of shock. Many reindeer and other wildlife, though, were killed. For those citizens that were around and not just in near proximity, a great, almost blinding flash was observed, with minor and much smaller debris from the object that didn't initially make it to ground level subsequently held up in the atmosphere, later burning up as meteors in the night sky. Right across Europe, over a sequence of nights, a night-time glow was seen, a visual reminder of what had transpired on the morning of June 5.

Nearer to the impact, reports emerged of people being physically knocked to the ground by the blast, one person being felled into a state of unconscious, only to awake several days later to a scene of devastation around him, where once were trees and vegetation. Reports also came to light of people being 'singed' by the intensity of the heat generated from the blast.

The blast equivalent of the Tunguska event is in line with an explosion creating 185 times more energy than the Hiroshima bomb. If the Tunguska fireball had occurred in a densely populated area the devastation would have been substantial, long-lasting, and costly, both in terms of life and rebuilding.

Strikes on Other Worlds

Collisions between bodies in the Solar System continued long after its formation, with the peak of activity occurring during an interval called Late Heavy Bombardment (LHB), around 4.1–3.8 billion years ago. The LHB period is thought to be the phase of evolution in the Solar System that was to eventually shape the terrestrial planets and moons. It could even be when life itself commenced, as planets gained atmospheres and other crucial elements that could well have established the premise for existence. The LHB period corresponds to the Neohadean and Eoarchean eras on Earth.

Our own Moon provides us with 'living' evidence of the continual bombardment of debris from space right up to the present day. It is believed that the Moon's crust formed around 3900 million years ago in a division of the Moon's evolution known as the pre-Nectarian period, with the next geological phase in lunar history, the Nectarian, in line with the LHB period. Pitted, cratered, and scarred, the Moon continues to be struck on a daily basis, but with its smaller mass it poses less of a target, and with no atmosphere to protect itself from even the smallest of objects that would readily burn up in our own atmosphere, it really is open season all year round for our lunar companion.

A lot of what has hit Earth has either fallen on wasteland or uninhabitable areas, or most likely in the sea, given the land-to-sea ratio. But there is one particular strike that continues to challenge the eye, the Barringer meteor crater in Arizona. The crater represents in plain view all that we associate with a direct hit: a substantial indentation on Earth, and testimony to what can and still could happen when all the other documented strikes on Earth are but confined to the textbooks, with no visible evidence. Here, some 50,000 years ago, relatively recent in Earth's history, an iron meteorite plowed into the ground. The impact from this 30-m traveler caused a crater 1.2 km wide, leaving a permanent scar on the planet's surface.

It is estimated that over 3000 meteorites measuring over 1 kg reach Earth every year with, to date, around 160 impact craters having been recognized, the Barringer crater being the biggest and most substantial. It is this crater that should serve as testimony to what such a relatively small body can inflict on Earth. As with the Tunguska event, the question of the level of destruction once again springs to mind, were the Barringer crater to be relocated into the middle of a city.

Because of the advancement in technology, we are being made acutely aware on a daily basis as to what bodies are out there, with regard to size, composition, and trajectories, even labeling them with names should their status and presence deserve such. For centuries these objects have been constant in Earth's evolution, even today wandering silently through space as citizens on Earth go about their daily business. Does it do us any good to know of their existence? Has it enhanced our understanding and made us more aware of our place in the universe? This type of question is only answerable by the individual and depends on how his or she perceives the universe.

Perhaps there is an order and a sequence to everything, one that was responsible for the eradication of a whole way of life for the dinosaurs, a total redirection for life on Earth. A substantial encounter with a large body may be purely random or part of a cycle, as we will discover later. Even so, all the dinosaurs could have done is watch and perish, their fate sealed from the moment that comet or asteroid sighted Earth. As helpless as they were, we virtually are, too.

Comet Shoemaker-Levy 9

However, what if we were to discover a comet, and watch it plow into the atmosphere of another planet in our Solar System? Would this not give us an invaluable insight into how such a collision on Earth may look to an observer watching us from another planet? Although there is little, or to be frank, nothing we could do to prevent a collision, surely a study of the effect would give a modern-day bearing on how these events occur? The opportunity to observe such an occurrence presented itself in 1994, when the remnants of Comet Shoemaker-Levy plowed into the largest planet in our Solar System, the mighty Jupiter.

The event, the first collision between two Solar System bodies ever predicted and observed, was watched with excitement by astronomers using telescopes across the globe. Out in orbit around Earth, the Hubble Space Telescope (HST), launched in 1990, turned its array of instruments towards Jupiter, including its 2.4-m Ritchey-Chretien reflector, unhindered by Earth's murky atmosphere. Hubble's instruments were able to analyze the collision in near-ultraviolet. But, even with the presence of the HST in the astronomer's armory to view the event, by a strange quirk of fate, NASA also had at its disposal the Galileo spacecraft, which was already on its way to Jupiter. Never in history had an event of this kind been so widely and closely observed, with so many varying types of scientific equipment (Fig. 7.2).

Fig. 7.2 A composite photograph assembled from separate images of Jupiter and Comet Shoemaker-Levy 9 (courtesy of NASA/ESA)

One very significant piece of information that would transpire from observations of Comet Shoemaker-Levy's encounter with Jupiter concerned the role of Jupiter in our Solar System. As we know, Jupiter has far-reaching gravitational fields that have an influence on many bodies that come close enough. It also transpired that Jupiter has the effect of reducing the amount of space debris that makes its way into the inner Solar System. The planet serves as a guardian against allowing larger threats to venture beyond it, towards Mars and Earth, by corralling, shepherding, and assisting in implementing an order from what could be chaos. Jupiter has a significant influence in this region of space, its mass imposing itself in many different ways, and one of the strangest of them all is the ability to almost filter and maneuver smaller objects out of harm's way, shielding the inner planets.

Discovered back on March 25, 1993, a previously unknown comet came into the view of Carolyn and Eugene M. Shoemaker, along with David Levy. Formally designated D/1993 F2, the comet was to become synonymous with the impact that would follow—that of Shoemaker-Levy 9 and its descent into the atmosphere of Jupiter. While conducting a program of observations designed to uncover NEOs, Shoemaker-Levy 9 was detected by the trio, who had previously discovered eight other periodic comets as a team.

This new find had been captured by the massive gravitational influence of Jupiter, and had subsequently fallen into a spiraling and decaying orbit around the planet. Discovered in photographs taken by the 18-inch Schmidt telescope at Palomar Observatory in California, Shoemaker-Levy 9 became the first comet to be observed orbiting a planet, with further study indicating that the comet may well have been caught by Jupiter's pull perhaps as long ago as 1929. That in itself proved an oddity. We know that cometary orbits often form around stars, but to be drawn into a planetary orbit was something quite special and unusual. This served to reinforce just how strong the gravitational fields are around Jupiter. Before the comet's eventual demise, it appears to have established itself in a two-year orbit around Jupiter, passing about 22,200 km above the planet's cloud tops on July 8, 1992.

However, these powerful fields were slowly but surely tearing Comet Shoemaker-Levy 9 apart, with the only possible outcome a headlong dive into the planet. The orbit of the Shoemaker-Levy 9 had brought the body within Jupiter's Roche limit, the minimum distance to which a body can approach its primary body without being torn apart by tidal forces.

The Roche limit, formulated by French mathematician Édouard Roche (1820–1833) in 1848, is sometimes referred to as the Roche radius. This is the distance within which a celestial body, held together only by its own gravity, will disintegrate due to a second celestial body's tidal forces exceeding the first body's own capability to be held together by its own gravity. In essence, Comet Shoemaker-Levy 9's own gravity was not strong enough to prevent the greater forces at work in Jupiter from tearing its nucleus apart.

The demise of the comet's nucleus became all too apparent over the coming months, with photographs revealing the true extent of its fragmentation. At least a dozen of the disintegrating comet's nuclei were to reveal themselves as the body approached its fate. The comet's nuclei lined themselves up like a glowing string of pearls around Jupiter. The original dozen became 21 separate fragments by mid to late July of 1994, as all resistance presented by the comet was lost. Thanks to the Sun's perturbations, this encounter was to be the last rendezvous with Jupiter, the Sun's influence having changed the comet's nearest approach to the planet (perijove) this time around to less than Jupiter's own radius.

Over the period of July 16 to July 22, 1994, the 21 fragments fell towards the planet. Following the practice of seeing previously observed comets fragment, the Shoemakers with colleague Levy labeled each individual piece of nuclei as "fragment A," "fragment B," and so on. The remnants of the nuclei struck the planet at intervals of seven to eight hours over the coming four days or so. Traveling at a speed of 220,000 km per hour, and ranging in size from a few hundred meters across to around just under 2 km wide, the descent into Jupiter's atmosphere began.

Each one of the fragments plowed into the unobservable night side of Jupiter, beyond and out of sight of Earth-based observers and the Hubble Space Telescope. Fortunately, though, NASA's Galileo probe came to the rescue, sending back impact images as each fragment exploded with tremendous energy into the Jovian atmosphere, creating a bubble of super-hot gas as it did so. These 'fireballs' would subsequently rise back out of Jupiter's atmosphere, depositing dark clouds of ejecta on top of the Jovian clouds.

Earth-based observers did not have to wait long to see what mark the comet had left on the planet. As Jupiter rotated (9.92 h rotational period), dark areas were clearly visible where the fragments had struck the planet. About one-third of the fragments produced little or no observable effects, suggesting that the fragments were relatively small in nature, perhaps less than 330 feet in diameter. Fragment G turned out to be the heaviest and largest of all the Shoemaker-Levy 9 debris deposited into Jupiter, with an estimated diameter of 1100–2000 feet. Its temporary legacy on the face of Jupiter was a multi-ringed black cloud, larger than Earth's diameter. Its impact was the equivalent of at least 48 billion tons of TNT exploding in the planet's atmosphere. The impacts remained on view for many weeks to follow, before slowly fading and eventually disappearing.

What if Comet Shoemaker-Levy had struck Earth instead of Jupiter? Don Yeomans of NASA's NEO program said the collision for Earth would be "catastrophic." Yeomans commented that "if something of similar size [to the size of Comet Shoemaker-Levy 9] hit Earth—we're talking about 2000 megatons of energy—there would be serious regional devastation or a tsunami if it hit the ocean."

Are There Civilizations Elsewhere?

Among the many and varied alternative explanations for great extinction events on Earth, is it possible that these have nothing whatsoever to do with rogue comets or asteroids? Could this global devastation actually be caused by a massive gamma ray burst?

These huge galaxy-wide bursts of electromagnetic radiation are caused by either the result of the dying explosion of a hypernova or when two neutron stars collide. Hypernovae seem to be the more likely of the two proposals, with a violent explosion hugely more aggressive than that of supernovae. Considered to be one of the potentially most devastating events in the universe, a hypernova would certainly account for a widespread annihilation of whatever life existed on planets within its reach. Lasting anywhere from a few milliseconds to several minutes, one such hypernova has the power within its gamma ray burst to wipe out life across huge swathes of the galaxy—a sort of cosmic clearing and leveling of the playing field.

Research has determined that the probability of Earth having been on the receiving end of one of these destructive gamma outbursts is 60%, given the time frame of a billion years in which to occur. That statistic rises to 90% when extending the parameters to the last 5 billion years. With this information, it would seem more than just a speculative guess that a gamma ray burst was responsible for the first great extinction event that took place in Earth's history over 440 million years ago.

This explanation would also go some of the way to explaining the Fermi paradox. Named after Italian physicist Enrico Fermi (1901–1954), together with American physicist and colleague Michael H. Hart, (1932–), he posed the question as to why, given the vastness of space and the possible evolution of life elsewhere, we haven't actually encountered any alien life form. One possible explanation could well be that of a gamma ray outburst. Our closest alien civilization was possibly wiped out by such a burst, resulting in the deafening silence that we've experienced on Earth since we started listening.

Fermi's classic statement "Where is everybody?" resonates with many who believe contact in some shape or form should have by now been established. With as many stars as there are in our galaxy, upwards of 400 billion, there are roughly an equal number of galaxies in the observable universe. In other words, for every star in our own Milky Way, there is a galaxy out in the universe to match it number-wise. However, absence of evidence is not evidence of absence.

As another analogy, for every grain of sand there is on every beach on Earth, there are 10,000 stars. Naturally, not all stars are classed as mainstream ones, like our own Sun. Given that, one has to also balance the debate by considering the likelihood of the evolution anywhere else of Earth-like planets. However, based on the likelihood that a very conservative 22% of stars are capable of having Earth-like planets evolving around them, there still remains the possibility of a potentially habitable Earth-like planet orbiting at least 1% of the total stars in the universe, a total of 100 trillion Earth-like planets.

One then has to ask about the level of technological advancement achieved by another civilization, and ask whether they are actually capable of interstellar communications. Simple radio waves are transmitted across Earth's globe on a daily basis, many continuing into space. SETI (Search for Extra-terrestrial

Intelligence) has been scanning the skies for decades in search of such a signal, but nothing has ever been received, that is, nothing that hasn't been disputed with a counter claim to what has actually been heard. The silence becomes more perplexing when you think about all the Sun-like stars that are in a more advanced stage of their lifespan compared than ours, meaning that many civilizations have already evolved while Earth was in its infancy. This in turn would suggest a potentially thriving universe.

Created by Russian astronomer and astrophysicist Nikolai Kardashev, the Kardashev scale is a method of measuring a civilization's level of technological advancement, based on the amount of energy that civilization is able to use for communication. The scale helps to group intelligent civilizations into three broad categories:

Type I: Civilization has the ability to use all of the energy on its planet. Also known as planetary civilization, a Type I civilization has harnessed the ability to use and store energy that reaches the planet from a host star.

Type II: Civilization can harness all of the energy from its host star. If this looks remarkably like a Type I civilization, one has to imagine a Dyson sphere, a hypothetical structure that allows for the capturing of a great percentage if not all of the host star's total energy output.

Type III: Civilization has achieved the Type II civilization state, but grown in its capabilities to enhance energy so much so that it now commands power comparable to that of an entire Milky Way Galaxy.

Since the three were formulated, other proposed levels of civilization have also been put forward to compliment Types I–III.

In essence, the Fermi paradox has no answer, only a host of possible explanations as to why we remain so alone. One explanation is that there are no signs of Type I or Type II civilizations because there are no such civilizations in existence, leaving everything on the same level as the civilization on Earth, a Type I. Just to add to this rather potentially startling realization that we really could be on our own is something called the Great Filter, a theory proposed by American economics professor Robin Hanson in 1998.

The Great Filter theory states that, at some point between pre-life to Type III intelligence, there is a 'wall' that all or nearly all attempts at life come up against. There's some point in the evolutionary process whereby the evolution of life literally hits a point that makes it extremely difficult if not impossible to progress beyond. This is the Great Filter.

The Great Filter would account for the equally Great Silence that we continue to encounter, and very neatly both explains and realizes the staggering importance of humankind's place in the universe. However, this in itself begs three further questions. Is humankind one of just a few select civilizations to make it beyond the Great Filter, meaning that we are basically not alone, but the chances of a another civilization being in our locale is remote, to say the least? Are we therefore the first to make it past the Great Filter, pioneers if you will, which in turn makes us unique and, more frighteningly, truly alone? Last but not least, are we yet to encounter the Great Filter, the possibility that it still lies ahead of us, with the further possibility of us not clearing the Great Filter?

Whichever stage humankind is reckoned to be at with regard to Great Filter, behind it or ahead of it, there is also the prospect that there is more than one Great Filter. Hanson defines his Great Filter theory as "the sum total of all of the obstacles that stand in the way of a simple dead planet (or similar sized material) proceeding to give rise to a cosmologically visible civilization."

Therefore, was the extinction of the dinosaurs, by whatever cause, a derailment in the process of the Great Filter, meaning that we had to start all over again? Was the extinction a freak episode that caused a tangent, sending Earth's evolution on a different path from what would have occurred had the dinosaurs not have been wiped out?

If the filter exists, there are indeed many trials and tribulations to pass through in order to get to a stage where one considers our own place in the universe, something scientists refer to as the "observation selection effect." This effect suggests that anyone who is pondering their own rarity in the universe is inherently part of an intelligent life, and, in analyzing their own place, the thoughts they ponder and conclusions they draw are identical.

The Fermi paradox, the Great Filter, and the Great Silence all have to afford adequate room for the Rare Earth Hypothesis (REH).

This is the suggestion that the emergence of life was extremely improbable and that humankind is a one-off. What has developed on Earth is a set of circumstances that is quite remarkable and unique, with many random factors converging to create a situation that couldn't possibly be duplicated anywhere else in the universe. Indeed, it could be the perfect storm, from which intelligent life formed to the point where our very existence is then scrutinized and challenged.

The REH was first proposed by geologist Peter Ward and astrobiologist Donald E. Brownlee, and draws on a much wider set of random events and happenings to suggest how life began on Earth—not just Earth's orbital position with regard to our Sun, atmospheric conditions, or the presence of necessary organisms, but a much wider series of conditions. These conditions include the positioning of our own Moon; the position of the planets that make up our Solar System and their own individual unique composition; and the positions of the stars.

The question of "When and where?" then is unanswerable. With so many random elements at work, who can determine which is right? Any action may be deemed an action that is part of the great filtration at work, another challenge for life to succeed under the most difficult of circumstances, another bridge to cross or mountain to climb. However, despite this apparent randomness, there remains a strange and perplexing order to the universe, a give and take, an inhalation and exhalation, a positive and negative.

8. Our Own Debris

Dawn of the Space Era

If having to contend with the threat of possible annihilation from asteroids and comets wasn't enough, our own self-made debris orbiting Earth now poses an equal if not greater threat.

Every satellite launched had its own purpose to fulfil, very costly, time-consuming and ultimately necessary in the advancement of utilizing space for our own purposes. Just as the land, sea and sky has been used to supply power and transportation, space was always the next logical step. But, wherever human life has gone, traces have been left behind, whether derelict buildings or vast swathes of rubbish floating out at sea. We've also managed to leave our mark on space—something else that will need clearing up.

The Space Age began on October 4, 1957, with the triumphant launch of *Sputnik 1*, the world's first artificial satellite. With other countries looking on with envious eyes, the former Soviet Union had achieved what others had been scrambling to do, placing hardware in space.

Whoever and whatever followed was to be but a bystander to the conqueror, the Soviet Union. Having had a taste of this new world of space exploration, global eyes now turned to the next big step: the Moon. The competitiveness that sprang forth, though shrouded in a political hue, like *Sputnik 1*, did of course have a positive aspect, in that space itself had taken on a different meaning for so many. Over the decades to follow, our lives would change because of our entry into this new dimension of exploration.

However, once this new era began, it would also herald the commencement of a countdown—a countdown that would ultimately lead to the inevitability that much of what was sent up would naturally have to come down. Was this an oversight at the time? Probably not. A lot of scientists probably had an inkling that,

© Springer International Publishing AG 2017   151
J. Powell, *Cosmic Debris*, Astronomers' Universe,
DOI 10.1007/978-3-319-51016-3_8

in time, space junk would litter our skies and, despite future planning that would hopefully see debris fall in unpopulated areas through controlled re-entry, there was always going to be a risk that whatever was planned wouldn't quite work out as hoped.

Like the sea lanes that traverse our oceans, our own 'ships' now occupy the space lanes above our atmosphere, progressive and necessary but, with only one in 10 satellites still operational by the end of 1990, a mounting problem that would only escalate in time. By the end of 2015, the number of satellites launched was estimated at 6600, of which 3600 remain in orbit with only about 1000 of them still operational. Therefore, with the balance in dead satellites measured at around 2000, and the likelihood that this number will steadily increase, the legacy of the early days of utilizing space is all too apparent. It is a mounting problem that future generations are unlikely to thank us for.

And there is lots of other space debris in orbit. A space census conducted as of midnight September 30, 1988, cataloged 7122 objects circling Earth, meaning that satellites account for only a proportion of what is in orbit, with other bits and pieces of hardware accounting for the rest. By 2013, it was estimated that a 170 million pieces of debris were in orbit around Earth, including some 29,000 objects larger than 10 cm in diameter. Only a very small proportion of these were actually being tracked, meaning that the likelihood of a fragment re-entering Earth's atmosphere at random was significant. It is estimated that approximately 17,000 human-made objects have re-entered Earth's atmosphere since 1957, many incinerated during passage. However, a proportion managed to survive re-entry, making landfall across the globe.

Down Comes the Debris

Generally, if you venture more than 1000 km above Earth's surface, the amount of satellite traffic decreases sharply. This initial area of space that starts at around 200 km up is known as low Earth orbit (LEO). The outer limit of LEO is around 19,300 km. The further out the satellite, the longer it takes to orbit Earth, and, at 965 km, one complete orbit takes around 104 min, compared to a satellite 160 km up, which orbits the globe in 88 min. Climb further to 20,000 km and an orbit takes around 12 h. During the 12 h (a half-day orbit), the satellite will track roughly the same pattern over the surface of Earth every day. Referred to a 'ground-track' the sequence continues day after day, with the satellite following a semi-synchronous orbit around the globe. Over time, the synchronization begins to erode somewhat, due to Earth first not quite rotating 360 degrees in a day and the fact that Earth bulges at the equator.

At 35,900 km lies a geostationary belt of satellites. Unlike the 12-hour semi-synchronous orbit, craft in this orbit travel at the same rate as the turning Earth. The semi-synchronous orbit rises and sets, but the geostationary orbit, which only works over the equator, keeps the satellite looking as if it were stationary in the sky.

In South Africa during September 1960, the remnants of what was believed to be an Atlas-Able expendable launch system came down. Numerous pieces of debris were reported, linked with an SM-65 Atlas missile. This particular method of launch lasted less than a year, between November 26, 1959 and December 15, 1960, before it was retired from service. Of the five Atlas-Able rockets built, two failed and the other three fell short of achieving orbit. Needless to say, with such a poor record let alone the cost involved the project was subsequently shut down.

Suborbital debris from the failure of a Thor booster used to launch the Transit IIIA satellite on November 30, 1962, fell to Earth in Cuba on November 1960. A motor and a propellant tank were among the reported debris. Thor was an American space launch vehicle whose previous incarnations mirrored those of the Atlas-Able, derived in this case from the PGM-17 Thor intermediate-range ballistic missile. Over time, the Thor family of rockets evolved into the Delta range.

No fewer than 11 pieces of stainless steel skin weighing 6–7 kg, along with a solitary rocket engine spherical pressure bottle weighing 48 kg, were collectively discovered in both Brazil and South Africa. The fragments were identified as debris from an Atlas booster for the Mercury MA-6 (23) *Friendship* 7 mission, which carried lone astronaut John H. Glenn, Jr., launched on February 20, 1962. The rocket was launched from Cape Canaveral with mission time lasting just under five hours. The purpose of

Glenn's flight was uncomplicated and very straightforward. First, it was to get Glenn safely into space, to observe his reactions to being in space, and to then return him safely to a point where he could be easily recovered. After completing three orbits of Earth, the craft came down 1287 km southeast of Bermuda, with Glenn alive and well.

In July 1962, part of a USAF satellite was discovered in Allegre, Brazil. The spherical pressure vessel measured 0.4 m in diameter. Launched on February 17, 1961, this was debris from Discoverer 20, also known as Corona 9014A, an optical reconnaissance satellite. For much of the mission the project went according to plan, but, towards the end of its time spent in space, Discoverer 20s film return capsule failed to separate from the main spacecraft, and all data was lost. The satellite's orbit decayed on July 28, 1962, subsequently falling to Earth as debris over Brazil.

Falling in a more populated area, a disc-shaped metal fragment weighing 9 kg fell on a street intersection in Manitowoc, Wisconsin. Identified as part of the Korabl-Sputnik 1, dubbed in the West as *Sputnik IV*. Launched on May 15, 1960, the fragment, measuring 0.15 m, plummeted to Earth on the morning of September 6, 1962, striking the middle section of the street on the corner of North 8th and Park. It was estimated that 24 pieces of *Sputnik IV* careened through the sky, with ear-witnesses stating a "thunder-like noise" accompanying the fragments. At the head of the debris was the largest chunk, which is believed to have landed in Lake Michigan.

Sputnik IV was part of the Soviet Vostok program (Korabl-Sputnik meaning 'ship satellite') and, although unmanned, was the precursor to the first human spaceflight mission, that of *Vostok 1*. However, the mission wasn't a great success. A computer bug in Sputnik's guidance system meant the craft was actually left pointing in the wrong direction, which consequently led to the satellite moving to a higher than desired orbit. However, the craft (piloted by a dummy cosmonaut) did send back vital data with regard to telemetry, and, with pre-recorded messages sent from the capsule, significant insight into radio communications between Earth and orbiting vehicles was obtained.

Fortunately, no incidents or accidents were reported because of the fall, although this did not spare the embarrassment the craft's demise caused to the Russian space authorities, who were forced to come and retrieve the piece of debris. The incident continues to be remembered and celebrated annually in Manitowoc with a fundraising event known as 'Sputnikfest,' which includes a Miss Space Debris pageant.

In October 1962, three segments of stainless steel skin (average size 0.9 × 1.2 m) and one piece of aluminum, still with a steel nut, bolt, and washer attached (about 5 × 5 cm, weighing 0.22 kg) were found on the Ivory Coast and Upper Volta. The pieces were identified as debris from an Atlas booster for the Mercury MA-8 mission, launched from Cape Canaveral on October 3, 1962. Lasting just over nine hours, the mission completed six orbits, with astronaut Walter Marty "Wally" Schirra (1923–2007) spending almost all of that time in a state of weightlessness.

On April 8 and June 28, 1963, respectively, two spherical pressure vessels were found near Broken Hill, New South Wales, Australia. The debris was linked with an Agena rocket, used to launch USAF test satellites. The pressure vessels were thought to be from two separate launches: December 14, 1962, and a later mission on January 7, 1963. During the same month, both stages had fallen to Earth, landing in the extreme northeast of southern Australia. The first fragment landed on Bollard's Lagoon Station, a cattle run about 160 km north of Boulia, in April; the second fragment landed about 80 km away, on Mount Surt Station in June. A third pressure vessel, referred to as a "ball" in several accounts, was also recovered at Muloorina on July 12, 1963. There are varying accounts of size, but it was generally agreed that the first object to fall in April weighed just over 5 kg and measured 35.5 cm in diameter. The second to fall was a slightly larger "ball" weighing 8 kg with a 41-cm diameter. The third fragment was a much smaller sphere, some 15 cm in diameter.

More stainless steel skin was discovered near Concordia, Argentina, in May 1963, weighing in at 2.7 kg. This piece of debris was identified as part of an Atlas booster for the Mercury MA-9 Faith 7 mission, launched on May 15, 1963. With a crew of one, Leroy Gordon "Gordo" Cooper (1927–2004), and space legend Rear Admiral Alan Bartlett "Al" Shepard Jr. (1923–1998) as stand-by, the capsule spent one day in orbit around Earth, traveling at a velocity of 17,500 mph (28,200 km/h) and completing just under 23 orbits before it re-entered.

May 1963 saw the recovery of a fragment believed to part of a Soviet spacecraft. The debris was found north of Pretoria, South Africa. It is possible that it was part of one of the Cosmos missions, although confirmation as to which one and launch date seems never to have been resolved.

In March 1964, a spherical object, metal in nature and weighing 11 kg, fell near Belem, Brazil. The fragment is thought to have been part of another Agena rocket. As debris is not guaranteed to fall soon after a launch, some fragments can never be clearly linked to a certain mission. This was also the case for what was believed to be debris of Soviet origin. Never officially identified, an area of British Columbia was showered with a number of fragments, also in March 1964.

Identified as pieces of a U.S. Department of Defense Titan IIIA booster stage, possibly a Trans-stage, launched on December 10, 1964, fragments from the booster were recovered in northern Argentina a day after re-entry on December 13, 1964. A metal sphere measuring 0.84 m in diameter, an aluminum cylinder (4 × 1.5 m) and four fragments of a rocket nozzle were found. This was the second test of this rocket system, the first Titan IIIA having failed on its maiden flight, after the Trans-stage refused to pressurize, resulting in engine failure and the subsequent inability to achieve orbit. The system had two further launches in 1965, but was subsequently retired. The 'Trans-stage' refers to the upper stage of the rocket, used on all Titan III models.

In January 1965, a piece of woven asbestos was recovered in Malawi. Its identity was never confirmed, but there was speculation it came from Kosmos 50. Launched on October 28, 1964, on an eight-day optical imaging mission, there was documented evidence of another fragment being found in Malawi on November 12, 1964. This fragment seems to have been more readily identified as connected with Kosmos 50, a satellite that at the end of mission was given an order to self-destruct. In total, 95 pieces of debris were catalogued, decaying from orbit between November 8 and November 17, 1964. This would account for the second fragment, but not the first.

Early 1965 also saw what appeared to be a "space-like" fragment wash up on the shores of Abaco Island in the Bahamas. The debris was thought to be a piece of an Atlas-Mariner I booster that had been deliberately destroyed for safety reasons shortly after its launch on July 22, 1962. The tiny island was to be nearly visited again by fragments from space, with the fairing from a Falcon 9 rocket floating to an equally tiny neighboring island, Elbow Cay. However, not to be outdone, Abaco Island also captured some fairing for itself, but no ordinary fairing. Originally thought to be from an Ariane 5 launch from French Guiana, this piece of debris actually turned out to be from a Mars program launch, destined to put the space rover Curiosity on Mars.

In June 1965, three pieces linked to the U.S. Department of Defense Titan IIIC development test launched on June 18, 1965, came down over several districts in India: Madiya, Pradesh, and Kota. In September 1965, a titanium sphere was recovered in Merkanooka, western Australia. Dubbed the Merkanooka ball, the sphere was later identified as a tank used for drinking water on the *Gemini V* spacecraft. Launched on August 21, 1965, from Kennedy Space Center with a crew of two, *Gemini V* had re-entered Earth's atmosphere on August 29, splashing down in the Atlantic Ocean.

Later on in the year, in December 1965, three metal spheres fell near Seville, Spain, believed to be parts from the Soviet *Luna 8* rocket. *Luna 8* was part of the Soviet Luna program with the objective of making soft landings on the surface of the Moon. *Luna 8*, though, suffered from a malfunction as it approached the lunar surface, a retrorocket failing to fire and causing the craft to crash. Launched on December 3, 1965, *Luna 8* (or *Lunik 8*) was the eleventh attempt made by the Soviets to place a craft on the Moon, ultimately resulting in failure.

Identified as part of a fragment from the launch of *Echo II* on January 25, 1964, a piece of plastic shroud was recovered in Australia. The *Echo II* spacecraft was essentially a 41-m balloon, made of aluminum Mylar foil laminate. Designed as a rigidized passive communications craft for testing tracking and communication techniques, this particular mission re-entered orbit on February 23, 1966.

In May 1966, an object was sighted by a Brazilian fishing boat while at sea off the coast of Brazil. The fragment, a helium pressure cylinder, measuring around 1 m and weighing 113 kg, was identified as debris from a stage of a Saturn rocket booster development test (SA-5) that had been launched on February 26, 1966, re-entering the atmosphere on April 30.

Also in May 1966, several pieces from the S-IVB stage of a Saturn development test rocket came down in the Rio Negro District of Brazil. In June a fragment fell in Columbia from an Atlas booster, thought to be associated with *Gemini 8*, the sixth manned spaceflight from this particular NASA program. On-board *Gemini 8* were Neil Armstrong and David Scott. Although Armstrong would later be the first to step on to the Moon, Scott piloted *Apollo 10* and was the first to drive a lunar rover, during the *Apollo 15* mission. Both Armstrong and Scott had a torrid time aboard *Gemini 8*, after the craft's thruster became stuck open, sending *Gemini 8* into a dangerous, whirling frenzy, from which, thankfully, both survived.

Another fall of debris from a later S-IVB was recorded in July 1966, with two fragments found in Peru and Zambia, respectively. Later that year, in October, a titanium pressure vessel from what was believed to be a Soviet craft fell on the city of Tomahawk, in the county of Lincoln, Wisconsin.

In January 1967, a metal sphere was found in Peru, with two spherical objects found in Mexico a month later. The first finding was identified as part of a Delta booster rocket, the second finding as debris from a Titan IIIC rocket. Another sphere came down in Mexico in July 1967, this one connected with the launch of the *Gemini 12* mission, the tenth and final mission of NASA's then current program. The Gemini series followed the Mercury missions, with the Apollo missions next in the sequence.

A sphere was also recovered in Saudi Arabia in September 1967, connected with the launch of *Explorer 35*, sent into space to make interplanetary studies. The craft dutifully went about its study for six years, before being switched off, after a mission duration of 2167 days, on June 24, 1973.

To end 1967, a piece of what was believed to be Soviet hardware fell in Finland during December.

Debris, understood to be parts of the lunar module descent stage of *Apollo 5*, was discovered in Columbia in February 1968, with more Soviet-linked hardware falling in the Gandaki Zone of

Nepal in March 1968. Following reports of a curious sighting over the Himalayas, four pieces of metal were recovered, with a link later established to a Kosmos 208 satellite, launched on March 21, 1968. In April 1968, a metal sphere was found in Mudgee, Australia, thought to be a pressure vessel from a Delta booster used to launch Biosatellite-II, an orbiting biological laboratory. Also in April 1968, the third stage of an *Apollo 6* booster was discovered in Angola. In August 1968 and September 1968, two separate incidents of spheres returning to Earth were reported, one discovered in Columbia, the other in Alaska.

Considering the amount of area in where debris could fall, the bulk of which the sea claims, fragments have been known to fall on decks of ships, the odds of which are literally astronomical. However, this was the case regarding a Japanese cargo ship, struck by several fragments while stationed off the port of De-Kastri in Russia. Considering the size of De-Kastri oil terminal, one of the biggest to deliver oil to the Asian markets, it seems even more improbable that the debris should single out this one particular Japanese freighter. Five sailors received injuries from the debris, with each individual fragment weighing around 10 kg.

If debris landing on a freighter is considered fluke enough, what then are the chances of a human being struck by a piece of debris? Lottie Williams of Tulsa, Oklahoma, therefore falls into a very elite class of people known to have been struck by debris. According to the Aerospace Center for Orbital and Re-entry Debris Studies (CORD), the odds of being struck by lightning are estimated at 1.4 million to one, with the odds of being hit by a fragment of space junk standing at one trillion to one. Fortunately, no serious injuries were sustained by either crew members of the Japanese cargo ship or Lottie Williams.

In July 1969, fragments from the first stage of a Saturn booster used to launch Apollo 2 on July 16, 1969, hurtled to Earth, littering the water and striking the deck of a German ship in the Atlantic Ocean. No injuries to crew members were reported.

A Soviet-linked metal pressure sphere plummeted to Earth in September 1969, striking the ground near Östersund in Sweden. Later that same year, in December, the shroud of an Atlas booster washed ashore near Marie Galante, Martinique.

In April 1970, what is believed to be part of a Soviet spacecraft was discovered in the West Cape area of South Africa. Thirty years later, in the same month, three fragments from a Delta II second-stage rocket crashed down at three separate locations across the same area of South Africa. In July of the same year, a spherical pressure vessel was recovered near Lai, Chad, with an assortment of debris falling on Kansas, Texas, and Oklahoma in August 1970. Both incidents have been linked with Soviet craft, although not with a great deal of certainty for the fragment that fell in Chad. The later finds across America were linked with *Kosmos 316*, a satellite launched in December 1969 from the Tyuratam Missile and Space Center in Kazakhstan. The satellite attracted much global attention because of the craft's unusual orbit, which finally decayed late in August. In March 1971, three separate spherical pressure vessels were discovered in North Dakota, with subsequent links being made to a U.S. launch.

Debris from the Soviet *Kosmos 482* satellite, launched on March 31, 1972, fell across New Zealand a month later. Four titanium spheres were found in an area near Ashburton, with a fifth discovered six years later in 1978 near Eiffelton. *Kosmos 482* was an interplanetary probe, with a mission to establish orbit around Venus. However, the probe, which followed the *Venera 8* probe launched four days earlier, never broke free from the orbit of Earth, with its descent back to Earth beginning just 48 h after lift-off.

On April 3, 1972, four red-hot titanium beach ball-sized spheres, weighing about 14 kg each, landed within a 16-km radius of each other, again near Ashburton, scorching holes in a field full of crops and making deep indentations in the soil. Space law requires that where space junk is found and subsequently linked to its owner, it is to then be returned. However, the owners, namely the Soviets, denied all knowledge and dismissed the link between the debris and it being that of *Kosmos 482*. Ownership then fell to the landowner. New Zealand South Island residents had reported rumblings and flashes of light in the sky several weeks before the actual spheres were discovered. The incident became the stuff of local legend, being referred to as the "Ashburton Balls Space Incident".

Falls of manmade debris across the United Kingdom are rare, but in May 1972 fragments from *Gambit 3* (Gambit mission 4435), a U.S. reconnaissance satellite, were discovered. Launched on May 20, 1972, from the Vandenberg U.S. Air Force base via a Titan III Agena D rocket, *Gambit 3* suffered a malfunction, sending it back to Earth. Officials did not expect any remains to make it through Earth's atmosphere intact, but they did, covering an 8-km radius, some 120 km north of London. The fragments included a spherical titanium pressure vessel, circuit boards, and pieces of glass.

Kosmos 954

Late January 1978 saw the demise of Soviet built satellite *Kosmos 954*. The satellite came down in northern Canada, spreading debris across a substantial area. Much of what came down was radioactive material from the spacecraft's nuclear power generator, with the debris field estimated to have covered literally thousands of square km. Canadian and American authorities mounted a search for the radioactive debris, but only a fraction of what is thought to have come down was ever recovered. The incident involving *Kosmos 954* was to spark international concerns over debris of this nature.

Satellite debris that falls to Earth and is solely comprised of inanimate material rarely stirs interest in the media, but *Kosmos 954* presented the global media with a sensation. For here, plunging back towards Earth, was a reactor core containing over 43 kg of highly enriched uranium, formed into carbide spherical projectiles and encased in carbon discs, not dissimilar to ice-hockey pucks. The reactor had also been functioning for over four months and, during that time of operation, generated fission products such as plutonium, cesium, and strontium. Furthermore, and to intensify the media frenzy, should one of these puck-shaped objects manage to survive re-entry and arrive intact on Earth's surface, the radiation level emitted from just one of them could kill at a distance of 1000 feet away.

Launched from Kazakhstan on September 18, 1977, *Kosmos 954* seemed to behave a bit abnormally from the outset as, shortly after takeoff, ground control staff wrestled with the trajectory of the spacecraft, with its erratic behavior causing the non-release of the nuclear power generator. Just over a month later, on October 29, the North American Aerospace Defense Command (NORAD) revealed that the satellite had strayed out of its intended orbit, with a prediction made for its return to Earth sometime during April 1978.

By December and under mounting pressure to act, the decision was taken by the U.S. National Security Council (NSC) to begin planning for the satellite's imminent uncontrolled re-entry back to Earth. Early 1978 saw *Kosmos 954*'s demise continue, with a loss of altitude and the craft's inevitable decay and descent. When it had become clear that *Kosmos 954* would fall back to Earth, the U.S. State Department decided that the best course of action would be to alert selected recipients of the impending situation, and on January 18, 1978, contacted via secret message its allies in NATO along with Australia, New Zealand, and Japan, where diplomats were made fully aware of what might transpire.

Despite Soviet officials attempt to instill calm over the escalating situation with claims that *Kosmos 954* would simply burn up upon re-entry, thus disposing of the threat, the wider audience was now all too aware of what some sections of the media were now referring to as the "killer satellite." Ultimately, Soviet officials were proved wrong, as *Kosmos 954* plowed straight through the atmosphere, plummeting back to Earth a month before NORAD's predicted return date in April. On January 24, 1978, with the world's media in attendance, *Kosmos 954* finally made landfall near Yellowknife, the capital city of Canada's Northwest Territories. Situated on the north shore of the Great Slave Lake and 400 km south of the Arctic Circle, Yellowknife witnessed the final moments of *Kosmos 954*, as satellite debris showered the icy tundra, covering thousands of sq. km.

Officials from both Canada and the United States then instigated Operation Morning Light, a five-month search of the area to recover as much of the debris as possible. With only a fraction ever recovered, history would subsequently confine *Kosmos 954* to the history books as a potentially life-threatening incident with an unremarkable end. In conclusion, all that investigators had was nothing more than a collection of fragments—no radioactive threat and, thankfully, no loss of life. A proportion of those involved in the investigation would, in due course, relate that the retrieval of debris from the *Kosmos 954* crash site had little to do with 'picking up the pieces' and more to do with finding out how far Soviet satellite technology had advanced.

Skylab

One of the most substantial pieces of hardware to fall from the skies was that of America's first manned space station, Skylab (Fig. 8.1).

Skylab was America's first and only solely U.S. space station and, at 77 tons (70,000 kg), made for a quite magnificent piece of technology for its day, an orbiting workshop for science-based experiments, the like of which had not been seen before. Skylab also marked a notable staging post in spaceflight advancement, for here the world now had a platform from which further, wider, and deeper projects into space could be planned and launched. A beachhead now existed in space where, for the first time, with longer duration spaceflights planned, the effects on humans of the exposure to prolonged periods of weightlessness could properly be

Fig. 8.1 As the crew of Skylab 2 departs, the gold sun shield covers the main portion of the space station. The solar array at the top was the one freed during a spacewalk. The four windmill-like solar arrays are attached to the Apollo telescope mount used for solar astronomy (courtesy of NASA)

studied. In turn, there was the possibility of perhaps one day reaching further afield than the Moon.

Although NASA's vision of such a cosmic steppingstone had now been realized, there was a fundamental and critical flaw in the planning, which, once the glory of attaining orbit and carrying out Skylab's purpose had been achieved, created far more headaches in the years that followed these glory days. What NASA had failed to do was look at the endgame scenario in the life of Skylab. NASA had estimated the lifespan of Skylab to be nine years but, with the cost of properly installing control and navigation systems that would allow Skylab to make a controlled re-entry after its time in space, this was a bridge that NASA knew was coming yet ultimately failed to prepare for.

Late in 1978, NASA's engineers discovered that Skylab was no longer maintaining its orbit; instead, decay had set into its orbital pattern, and the research laboratory was now beginning to spiral downwards. With global concern now being voiced by various governments, at the forefront of which was the risk to life as 77 tons worth of metal could potentially plummit back to Earth, NASA responded with a plan of action.

Currently in development was the first reusable space vehicle, the space shuttle, which, during its lifetime that was then unforeseen, would revolutionize space travel, although sadly at a human cost. The plan entailed speeding up and completing development of the space shuttle so that when it was eventually launched it could be used to maneuver Skylab into a higher orbit, extending the life of the space station by perhaps as much as five years. With this new higher orbit achieved, Skylab would then pose no threat and would simply, at the end of its life, join the other millions of tons of space junk. A seemingly feasible and credible plan, this was duly thwarted over the coming months with funding and other issues delaying the development of the space shuttle, and thus consigning Skylab to its ultimate fate.

With Skylab's fate now sealed, somewhat perversely, but in the view of some not inappropriate, Skylab-inspired parties were held across America, seemingly in a proportion of cases mocking NASA's inability to say exactly where and when Skylab would eventually come down. It rapidly became an entrepreneurs' market, with one hotel holding a poolside disco party, and a target sign for Skylab ready for its crash landing on Earth. Elsewhere, you could purchase a can of "Skylab repellent" or wear a T-shirt with a large bullseye emblazoned on it. This heightened state was not just confined to the United States, although, in other countries, the potential impending crash of Skylab was taken more seriously, with officials in Belgium planning to sound air raid-type sirens in the event of Skylab missing its anticipated 7400-km estimated field of debris across the Indian Ocean and Australia.

On July 11, 1979, Skylab finally began its descent through Earth's atmosphere, with NASA engineers firing the station's booster in an attempt to send the laboratory into a tumble that would see it eventually end up in the Indian Ocean. The world braced itself as Skylab plummeted to Earth and, for the most part, the booster firing seemed to work, sending a significant amount of debris into the ocean, with other debris crashing down over populated areas of western Australia although, thankfully, with no reports of injuries.

As Skylab had loomed ever closer, the American media fielded a circus at NASA's expense, offering "Skylab insurance" should anyone be injured or indeed killed by a piece of debris from the laboratory. One such jape was to ultimately backfire, thanks to the ingenuity of one person. The San Francisco Examiner had wagered $10,000 to the first person to arrive at their office with debris within 72 h of Skylab crashing back to Earth. The newspaper thought it would be on relatively safe ground, given the fact that NASA continued to predict the same crash zone as it had been predicting for weeks, well outside of American soil. When Skylab debris finally rained down on Australia, pelting the house of a young teenager, one Stanley Thornton, he quickly collected some of the pieces and, with not a moment to lose, took an airplane to the United States to claim his prize, which in all fairness the newspaper duly paid.

Stan Thornton heralded from the port town of Esperance, and it is there that much of debris from Skylab went on show, at the Esperance Municipal Museum. With a large Skylab model awaiting tourists at the museum's entrance, visitors can see just how much of the laboratory didn't fall in the Indian Ocean and actually made it onto dry land. Among the displays are two chunks of a Skylab oxygen tank. Other museums in Australia also house fragments of

Skylab. The Balladonia Museum has as one of its displays a pair of large sheet metal pieces, one sheet bearing the name "SKYLAB" in large red letters, the other labeled "Airlock/Danger".

Salyut 7

Identified as debris from the Soviet Salyut 7 Space Station launched on April 19, 1982, numerous fragments were recovered around the town of Capitan Bermudez, Argentina, in February 1991. By the time Salyut 7 fell to Earth, it still had attached to it *Kosmos 1686*, one of the many craft that had been sent to join Salyut 7 during its marathon stay in space. It was hoped that *Kosmos 1686* could use its boosters to push the space station into a higher orbit, thus prolonging the space station's residency in space by an extra three years. It had also been hoped that by this time Buran, the equivalent of the U.S. space shuttle, would be in service, with this reusable vehicle being able to dock with Salyut 7. However, Buran was not to be at this time and, sadly for the Russians, it was a project that was ultimately shelved. In the end, unexpectedly high solar activity and increased atmospheric drag on the station finally brought its life in space to an end.

Salyut 7 was the last Salyut mission, by far the heaviest in comparison to its predecessors. It was regarded as the transitional phase link from the old style of space station to the new ones, with the up and coming Mir eradicating so many of the problems that had dogged Soviet stations for years previously, issues that were still all too apparent on Salyut 7 through power losses and leaks. Salyut 6 and Salyut 7 were very similar, with virtually the same equipment, almost a mirror image of the station's capabilities. Indeed, Salyut 7 was the back-up for Salyut 6, but with the Mir space station project overrunning its intended launch date, Salyut 7 was dispatched into space.

By the end of 1986, and with Mir now launched, equipment was transferred from Salyut 7 to Mir. Salyut 7 rapidly became a massive orbiting empty container, but, with the failure to prolong its life in space following the docking of *Kosmos 1686*, Soviet officials knew the fate of the space station was sealed. Salyut 7 finally fell from orbit, re-entering Earth's atmosphere on February

7, 1991. Unlike former Salyut re-entries, this was completely uncontrolled. With a combined weight of 40 tons between Salyut 7 and *Kosmos 1686*, the pair plummeted through the atmosphere, overshooting the intended resting place in the Pacific Ocean, with fragments showering the town of Capitan Bermudez, Argentina.

As the mass of twisted metal sped through the atmosphere, Argentina was treated to an impressive light show of flaming debris, with the sky became ablaze with hundreds of incandescent chunks of material illuminating the heavens as the remains of the space station swiftly moved southwest to northeast across the heavens. As dawn broke, residents scoured the area, finding what appeared to be distinct groupings of debris. Fortunately, there was no loss of life, and no damage to property was reported.

As for the Russian-made reusable space vehicle Buran (meaning "snowstorm"), the craft flew just once before being unceremoniously mothballed and literally left to rot. Buran performed 24 test flights, 15 of which were fully automatic, as part of its preparation for actual spaceflight.

The first launch attempt took place on October 29, 1988, but ended with a mechanical failure, when a platform next to the Energiya's rocket that was carrying Buran did not retract quickly enough. The rocket's computers subsequently halted the countdown. However, on November 15, 1988, Buran achieved entry into space for its first and, as it would transpire, last time, with all systems apparently functioning well. After two nearly circular orbits, and just an hour after its early morning launch window, Buran returned to Earth just one second earlier than planned. However, the craft was never to fly again. Unlike NASA, different 'space factions' were at odds in Russia, which simply meant no future agreement could be reached on Buran for what, at the time, both visually and possibly mechanically was a shuttle that could rival that of NASA's own creation.

One in a Trillion

On January 22, 1997, Lottie Williams was out walking her dog with friends in the early hours of the morning, around 03:30. Lottie, from Tulsa, Oklahoma, while looking to the heavens, witnessed a streak of light blazing across the sky, followed minutes later by the sensation of feeling that something had touched her shoulder. What she had sighted and indeed felt was part of a U.S. Delta II rocket, launched back in 1996. This brush with space junk is the only recorded account of anyone ever having been 'hit' by such debris, bar the five sailors aboard the Japanese cargo ship mentioned earlier.

A much larger and more significant part of the Delta II rocket, weighing 263 kg, came back to Earth with a distinct thud, crashing into the ground in Texas, narrowly missing an occupied farmhouse. This piece of debris turned out to be a large stainless steel fuel tank from the rocket, which had deposited itself in the front yard of a farm near Georgetown. Further away, another piece of debris from the Delta II embedded itself in a field in Seguin. The titanium sphere, virtually intact, was found halfway lodged into the ground.

The sensation Lottie had felt was that of a charred piece of woven material that hit the ground, according to her, "with a metallic sound." Speaking to a reporter from *Tulsa World*, Williams concluded that there was a connection between what she had witnessed in the sky and what she had felt, describing the object in the early morning sky as being like "a big, bright light, like a fire," with the object coming in her general direction. Williams continued by commenting to the reporter that "two sparks shot from the fireball and disappeared over a building".

On April 20, 2000, a portion of a different Delta II rocket re-entered the atmosphere over the Atlantic Ocean. Three objects were subsequently recovered in South Africa following the event, these being a stainless steel propellant tank weighing 260 kg; a titanium sphere weighing 33 kg; and a tapered cylinder that served as part of the main engine nozzle assembly, weighing 66 kg.

Costing $617 million, and weighing in at 17 tons, the Compton Gamma Ray Observatory met a fiery end on June 4, 2000, in a controlled de-orbiting maneuver that saw this heavyweight satellite, the size of a bus, splash down in the Pacific Ocean, southeast of Hawaii. Launched on April 5, 1991, the observatory spent an impressive nine years gathering data in space, completing over 51,600 orbits around Earth before a crippled gyroscope forced NASA to bring it down. Had NASA not taken the decision to do so, it had been calculated that a natural descent and burn-up would still have statistically left a 1-in-1000 chance of killing someone on Earth. Engineers at the Goddard Space Center subsequently sent radio signals to the observatory, inducing a series of rocket firing sequences that altered its trajectory, forcing the satellite into Earth's atmosphere.

During its descent into Earth's atmosphere, it is believed that the observatory's solar panels were the first to be lost, with subsequent parts being super-heated and burned away as it traveled down towards the sea below. Shipping and aircraft had been warned to stay well clear of the area, as some 6000 kg of debris plunged into the ocean.

The satellite's time in space had been a great success, being the first major space observatory to study invisible gamma rays, with the detection of over 2600 gamma ray bursts, and discover hundreds of unknown gamma ray sources. The observatory literally changed the way astronomers look at the universe.

In October 2000, a metal fragment was found in Wichita, Kansas, thought to be part of the fourth-stage casing from a Russian Proton booster, used to launch three Glonass navigation satellites. These satellites, of which many have been launched, offer the Russian Aerospace Defense Forces a distinct alternative to GPS, operating across the globe on three orbital planes.

January 2001 saw a third-stage component of a Delta II rocket re-enter the atmosphere over the Middle East. The titanium motor casing of the third stage, known as a PAM-D (Payload Assist Module-Delta), finally landed in Saudi Arabia. Weighing about 68 kg, the debris fell some 240 km away from the Saudi capital of Riyadh. After eight years in orbit, it was fortunate that the casing, found still displaying a Boeing part number, fell on such a sparsely populated area.

The Mir Space Station

The Mir space station was one of the most substantial pieces of space junk to fall back to Earth. Launched on February 20, 1986, Mir became a legend in its own lifetime, reflecting much of what Russia had already contributed to space and paving the way for not

Fig. 8.2 The space shuttle Atlantis docks with the Mir space station, June 1995. Part of the shuttle-Mir program (courtesy of NASA)

just Russia but other nations to embrace space as part of an accessible new world (Fig. 8.2).

Mir translates into English as "world" or "peace." However, in Soviet history, following the Edict of Emancipation in 1861, the word *Mir* referred to a much earthier and relevant statement of the time, that of a Russian peasant community, a 'society,' and the community's ownership of its own land. A popular Russian name is Vladimir, which means "the one who owns the world." This is derived from two Russian words: 'Vladet' (to possess) and 'Mir' (the world). However, with regard to the peasant communities, a system of state-owned collective farms replaced the Mir concept after the Russian revolution of 1917.

During its time in orbit around Earth, many other different names were given to Mir, everything from the insulting name calling of a "derelict" to the glittering accolade of "a marvel" and to the downright absurdity of "a lemon".

Whichever name was used, Mir had an irrefutable track record that saw it spend 15 years in space, three times beyond its planned lifetime, indeed, outlasting the Soviet Union that launched it into space in the first place. During its lifetime, the station hosted 125 cosmonauts and astronauts from no fewer than 12 different nations and, while in orbit, supported 17 space expeditions, conducted 23,000 scientific and medical experiments, and played host to 28 long-term crews. The space shuttle docked nine times with Mir that, as one of the greatest collaborations in space, showed the world that, in order to fully engage in the conquest of space, a global concentration of effort through international cooperation was the only way forward (Fig. 8.3).

Traveling at an average speed of 28,783 km/h, the 100-ton station orbited at a distance of around 400 km. Mir raised the first crops in space, enjoying both great new landmarks in the history of spaceflight but also great lows, indeed, the very depths of despair, at one time narrowly avoiding a catastrophic collision.

However, Mir was to finally be overshadowed as its reign came to an end at the dawn of a new era, with the arrival of the International Space Station (ISS). The arrival of the ISS drew much

Fig. 8.3 Photograph of the space shuttle Atlantis taken from the Mir space station during operations on March 23, 1996 (courtesy of NASA)

of the finance away from Mir, with Russian officials realizing that the future lay with the ISS and not the ailing Mir, and that perhaps it was time to bring the curtain down on the legend. Despite a rearguard action by those officials close to the project, the fate of Mir was therefore subsequently sealed, with the decision taken to de-orbit the station. In such a low orbit and without repeated booster interaction to keep Mir in a reasonably stable orbit anyway, this was not hard to achieve, but the feeling of loss hit hard upon the collective Russian conscience, after so much work, effort, time, and money had been spent to keep Mir up in space.

In the event that Mir overshot its designated crash site, perhaps landing in Japan or possibly South America, Moscow took out a £138 million insurance policy just to be on the safe side. Japan kept a close watch on the Mir space station as its orbit decayed further, and probably rightly so, as during its 86,000 orbits over the years, Mir's path had crossed over nearly every city on Earth and, with the country in line for an overshoot, officials in Japan had become worried.

Mir began its descent in Earth's atmosphere on March 23, 2001, re-entering above the Pacific Ocean near Fiji. Though most of the 130- to 140-ton (130,000–140,000 kg) station burned up in the atmosphere, a fair amount of debris survived, with an estimated 1500 fragments reaching the surface of Earth. In a reportedly quite spectacular smoke and light show, akin to a blaze of shooting stars, Mir plunged into the South Pacific, with the debris falling over an unpopulated patch of the ocean southwest of the British Pitcairn Islands.

UARS

NASA's Upper Atmosphere Research Satellite (UARS) hurtled back to Earth on September 23, 2001. Successfully deployed by the space shuttle *Discovery* on September 15, 1991, this 7-ton scientific craft orbited Earth 78,000 times, collecting data on chemicals in the planet's atmosphere. UARS made significant studies on carbon dioxide, ozone, chlorine, methane, nitrogen oxides, and chlorofluorocarbons as it conducted experiments into Earth's atmosphere and its interactions with the Sun. On December 14,

2005, UARS ceased operation but continued to orbit for another six years before finally returning to Earth.

Although NASA predicted that much of UARS would burn up in the atmosphere upon re-entry, some debris would survive, with 26 satellite components, weighing a total of around 545 kg, potentially making it back the surface of Earth. With a debris zone covering an 800-km area, UARS entered the atmosphere above American Samoa, with disintegration and subsequent fragments finally beginning to strike an area in the Pacific Ocean southwest of Christmas Island, about 480 km beyond the satellite's initial entry point. In total, it was believed that about two dozen metal pieces fell over the calculated debris field.

However, the final resting place for the debris had been the center of much debate before UARS had actually come down. The concern centered on just how the calculations for the satellite's entry and debris field had been formulated. The previous debris crash zone had been predicted across a large swathe of north-western North America, before being updated by the U.S. Air Force and subsequently NASA, placing the crash zone thousands of km away from its previously predicted crash sight. With such a shift of crash sites, this unnerved many observers, wondering if the site would be moved again.

The odds of a piece of debris striking an individual are one thing, but for a house in Uganda, debris from a European Space Agency (ESA) third stage of an Ariane 3 booster scored a direct hit. Falling to Earth in March 2002, a titanium pressure sphere from the booster, used to launch Gstar1 and Telecom 1B, landed on a home in Kasambya, Uganda. Luckily, no one was in residence at the time. In August 2002, more ESA debris returned to Earth, also from the third stage of an Ariane booster, this one though a later model, the Ariane 4. This particular sphere landed near the village of Manzau, Angola.

Tragically, February 1, 2003, saw the space shuttle *Columbia* disaster, where seven lives were lost. The debris from the accident covered 45,000 km^2 across eastern Texas and western Louisiana. Over the months to follow, over 80,000 pieces of wreckage were retrieved, being placed into storage for future analysis.

In April 2003, a sphere landed on farm near Mataquesquintla, Guatemala, later identified as debris from the Centaur stage of an

Atlas IIAS booster, used to launch Intelsat 806, a communications satellite. In July 2004, debris from the second stage of a Delta II booster fell over two different areas of Brazil, first near Cabeca da Vaca and then Batalha. The booster was part of the rocket that launched Mars Exploration Rover B (Opportunity) on July 8, 2003. Costing $400 million, Opportunity was launched from Cape Canaveral, landing at its target site on Mars, Meridiana Planum, on January 25, 2004, three weeks after its twin craft, Spirit, had touched down.

A titanium rocket-motor casing was found near Bangkok, Thailand, in January 2005. The casing (with a diameter of 1 m and a length of 2 m) was identified as fragments from the third stage of a Delta II booster, used to launch GPS IIR-6 on November 10, 2000.

A sphere was also discovered on a farm near Montividiu, Brazil, in March 2008, identified as debris from the Centaur stage of an Atlas V booster, used to launch WGS-2 (F1) satellite on October 10, 2007. A high capacity satellite communications system satellite, this craft was sent up for use by the U.S., Canadian, and Australian governments.

In July 2008, a metal rocket motor casing was discovered in Australia, identified as debris from the third stage of a Delta II rocket used to launch INSAT-1D. A steel propellant tank and two titanium pressure spheres were discovered in Mongolia, identified as fragments from the second stage of a Delta II booster, launched on September 25, 2009, and re-entering orbit on February 19, 2010.

Testifying to the notion that a proportion of hardware that goes up doesn't necessarily return to Earth within a short span of time are the remnants from the third stage of a GSLV booster. Used to launch INSAT-4CR, several metal objects recovered in Malawi in February 2011 were thought to probably be fragments of the booster. Despite being launched back on September 2, 2007, this actual fragment didn't re-enter Earth's atmosphere until March 3, 2011. A titanium rocket-motor casing was discovered in March of the same year, identified as debris from the third stage of a Delta II booster, used to launch GPS IIR-10 on December 21, 2003.

Also in March 2011, a hiker in northwestern Colorado sighted an object resting in a crater. The object, spherical in shape, was warm to the touch. Having contacted military aerospace officials, the hiker was initially told to contact the local authorities, in this case the local sheriff. However, after continuing to pursue the issue with higher officialdom, the hiker finally managed to get through to the NASA office that tracks space debris. The persistence of this individual is to be credited, as the object turned out to be a spent tank from a Russian Zenit-3 rocket, launched two months previously.

Launched on June 1, 1990, the Roentgen Satellite (ROSAT), named after German engineer and physicist Wilhelm Roentgen (1845–1923), went into orbit around Earth, with its mission to perform the first all-sky survey of X-ray sources, using an imaging telescope. ROSAT, a German Aerospace Center-led mission, was developed through a cooperative program between Germany, the United Kingdom and the United States and, during its time in orbit, produced a catalog of more than 100,000 X-ray sources. During its eight-year mission, which eventually involved more than 4000 scientists from 24 countries who were given access to monitor observations, ROSAT also discovered that, rather surprisingly, comets, too, emit X-ray radiation.

After being switched off on February 12, 1999, ROSAT finally de-orbited on October 25, 2011, with officials at the German Space Agency calculating that some 30 pieces of the satellite would probably survive re-entry. Officials estimated that the debris field for the satellite would cover an 80-km track of Earth's surface. On October 23, 2011, ROSAT's very productive and fruitful life came to an end, falling from space to a watery grave over the Bay of Bengal in the northeastern Indian Ocean.

More debris was to fall on a house in 2011, after a Russian communications satellite, Meridian 5, failed to achieve orbit, crashing back down to Earth in Siberia soon after its launch in December. One of the satellite's titanium fragments was reported to have made a 1-m hole in the roof of a house, ironically on a street named Cosmonaut. Thankfully, nobody was injured. Launched into orbit by a Soyuz 2.1B rocket from the Plesetsk Cosmodrome, problems with the third stage of the vehicle occurred, with an emergency command to shut the engines down sent from ground-based officials.

In 2012, a sphere was discovered near Mata Roma, Brazil, being part of a helium pressure tank from the third stage of an ESA

Ariane 4 booster. The booster was used to launch communication satellites Thaicom 3 and B-Sat1a on April 16, 1997. This particular piece of debris spent five years in space before re-entering Earth's atmosphere on February 22, 2012.

In March 2013, two metallic spheres, later identified as helium tanks from the third stage of a Long March 4B booster, fell to Earth near Buna, Texas. One of the spheres was discovered in a pasture used for grazing cattle. In July of the same year, a propellant tank was found near Ngezi, Zimbabwe, believed to the second stage of a Delta booster used to launch Symphonie 2 from the Kennedy Space Center, August 27, 1975. The tank spent 38 years in orbit around Earth before re-entering on July 14, 2013. The Symphonie range satellites were the first communications satellites built in a joint effort between France and Germany.

In December 2014 and January 2015, three small cylindrical tanks and a metal ring were discovered near Santa Rita do Pardo, Brazil. This collection of debris was thought to be fragments from a second stage Falcon 9 booster used to launch communications satellite Asia Sat 6 on September 7, 2014. In January 2016, two small spherical tanks were found in the Tuyen Quang Province of Vietnam, probably from the second stage of an SL-23 booster used to launch Elektro-L2, a weather satellite, on December 11, 2015.

Miscellaneous Debris

The term 'space junk' is not just confined to redundant satellites, but to all other home-grown pieces of debris, either accidentally misplaced in space or intentionally set free.

As far back as the 1960s, miscellaneous bits and pieces have been boosted into orbit around Earth. Take, for example, a glove belonging to Edward H. White II (1930–1967), the very first American spacewalker. While venturing outside of his *Gemini 4* flight in 1965, the glove parted company from White, making many circuits of Earth before, after around 30 days' worth of traveling around the globe, burning up in the atmosphere. *Gemini 4* was the second manned spaceflight in NASA's Project Gemini series and comprised a four-day mission carrying astronauts James A. McDivitt and Edward H. White II. This and subsequent missions were the forerunners in developing the technology and accumulative time spent in space by its astronauts that would eventually fulfil the dreams of President John F. Kennedy, and his vision to have a man on the Moon by 1970.

In March 2001, during a spacewalk to mount important equipment to the International Space Station, a foot attachment that was used to anchor spacewalkers to the end of a space shuttle's robotic arm, in this case the space shuttle *Discovery*, floated away from astronaut Jim Voss. Compared to some of the debris that is known to still be in space or known to have fallen, you would not consider this loss to be of any particular significance. However, space debris of any size is a problem when you are actually in space itself. As a consequence of losing the foot attachment, *Discovery* was forced to alter its orbital altitude, in order to circumvent meeting the attachment on the next orbit.

Among items lost forever are a spatula, accidentally misplaced by astronauts Piers Sellers and Michael Fossum while busy spreading a "sticky-like" substance to test out potential repairs for damaged heat shield tiles. Following the *Columbia* space shuttle disaster in 2003, this 2006 mission on board sister ship *Discovery* saw a sequence of tests to try out new safety techniques while under the restricted conditions of space.

Astronaut waste from over the decades of spaceflight has also formed part of the collective debris, with accumulations of urine filled bags still in orbit. These bags, which quickly freeze in the cold vacuum of space after being jettisoned, have accumulated over the years, although advancement in technology has in time allowed for the recycling of such waste, converting the urine into drinking water, thus dispensing with the need to jettison it into space.

A camera and a pliers were lost in 2007, the former courtesy of astronaut Suni Williams, the latter courtesy of astronaut Scott Parazynski, whose pliers were last seen floating away below the International Space Station. Also in 2007, NASA instructed its astronauts to deliberately ditch an unneeded 1400-pound (635-kg) tank full of ammonia. The tank had been part of the ISS's cooling system but, following an upgrade, was no longer required and was thus literally thrown into space, as it was deemed to be taking up too much cargo room. The tank spent over a year orbiting Earth before burning up in the atmosphere over the South Pacific Ocean.

One of the most expensive pieces of equipment to be lost was a $100,000 tool bag, accidentally let go by astronaut Heide Stefanyshyn-Piper during a November 2008 spacewalk to repair a jammed solar panel on the ISS. The bag, weighing 13 kg, was tracked by amateur astronomers as it circled Earth.

Stefanyshyn-Piper's bag makes for somewhat comical analysis in the world of cosmic debris. For within the astronaut's bag, among a number of grease guns and scraper tools, were another series of bags, taken up to the ISS in order to collect space debris!

9. Observing Meteors and Meteor Showers

Meteors, Fireballs, and Bolides

The observation of cosmic debris allows for both amateur and professional involvement. Not only this, but with so many eyes both human and electronic, scanning the heavens there has never been a more productive time for cataloging, tracking, and making new finds.

Apart from the Moon, planets, and stars, one of the first instigators of intrigue about the night sky can often be the glimpse of a shooting star—a meteor. The meteor's sudden and often momentary appearance has generated many a gasp of wonderment and, for those who subsequently turned to astronomy in whatever capacity, marked the start of a journey of discovery. For not only does the inquisitive mind ask for an explanation as to what has just been witnessed, but it questions further as to the meteor's origin and why, at that particular time, its life became vaporized in a flash of light.

Meteors are a great starting point for the amateur, for not only is every night a chance to see one, but the actual involvement level is a minimal one. All that is required is a clear, preferably Moonless sky and a little bit of patience. The biggest draw in observing meteors is that you do not require a pair of binoculars or a telescope, just your eyes, and preferably a good vantage point that allows you the best view possible of your horizons. Try and find an observing site that is also as dark as possible, away from artificial lighting, which can hamper observations and blot out fainter meteors.

So, what are we seeing? Luckily, unlike the pounding our own Moon has to withstand, Earth's atmosphere burns up the vast majority of smaller objects that come into contact with it. Anything from a grain of sand to a small stone or pebble (1 mm to

© Springer International Publishing AG 2017
J. Powell, *Cosmic Debris*, Astronomers' Universe,
DOI 10.1007/978-3-319-51016-3_9

1 cm) will burn up on entry, friction with air molecules making the body glow as its speeds through the atmosphere, generating that flash of light before disintegration. The larger lumps of rock can and often do make the passage through our atmosphere and, upon making ground level, are re-classified as meteorites.

Every day, literally thousands of small bodies rain down on Earth, with an estimate of one per second burning up in the atmosphere.

Where do they come from? There are many such pieces of debris orbiting Earth; many are fragments left over from the breakup of a larger parent body such as an asteroid or a comet, in an occurrence that took place possibly thousands or even millions of years ago. Following the breakup, the fragments have established an orbit around the Sun, into which Earth subsequently travels through in its own orbit. Whereas the debris from an asteroid is unlikely to be added to as the parent body is destroyed, a comet with a steady orbit around the Sun will continue to add new debris to what has already been deposited, constantly feeding the fragments that Earth passes through.

Throughout the year, apart from the more random debris elements, Earth also passes through more 'organized' debris, and, as it does so, a shower or meteors can be expected. It is during these occasions that anyone wanting to see more than just the odd meteor should prepare for a veritable feast. It is also during these annual outbursts that the observer can gain a better understanding of the types of meteors that can be seen, the varying brightness, the different trails, known as 'trains,' left behind, and in some cases, as with fireballs, accompanying sound effects as well!

Fireballs are less common than meteors, around one per hour on average, measuring more than a meteor, in excess of 1–1.5 cm. Because of their slightly larger size, the atmosphere has a little more to deal with, so the end result for a fireball is a more aggressive and ultimately more glorious burn, far brighter in appearance. Also, because some of these fireball objects have their rock-like surface pitted, combined with traces of other substances in their makeup, there could well be a distinct color to the fireball or, in some cases, a whistling noise accompanying the fireball's final moments.

Next on the scale working up from meteors, then fireballs, are bolides. Bolides are again a much larger body, possibly measuring up to a meter in diameter but, unlike meteors and, to a lesser extent, fireballs, are a lot less likely to be witnessed, with frequency estimated at one per day. These too are mostly dealt with by Earth's atmosphere but with an amplification of what you would see with a fireball. This amplification means that more people are likely to witness a bolide, even if they are not watching the skies with a purpose. Therefore, there is a more noticeable widespread reporting of such a phenomenon, as its final dramatic moments are in more recent times being readily caught on the many private and commercial cameras in circulation, and also cameras that are mounted on car dashboards. Beyond the bolide is very different territory indeed, for these monster-like objects lie at the business end of the scale, where very few fail to spot an incoming asteroid or comet.

So, without optical aid, the observation of meteors makes for a perfect first step into not only this field of observation but astronomy as a whole, as once the observer has an understanding of what is being seen (hopefully without some of the magic being taken out), they can perhaps turn a passing interest into an ongoing pursuit.

Annual Meteor Showers

On a night where the observer is just on the lookout for sporadic meteors, there may well be a reward for patience with five to ten meteors, given the amount of time spent watching the sky. Granted, this does not sound an awful lot but, throughout the year, there are times when annual displays of meteors provide a much richer bounty for the willing and the patient.

These displays will appear to come from one distinct radiant, a point of entry in the sky as Earth passes through a dense concentration of debris. These annual bursts in showers of meteors are thus named after the constellations in which the radiant is situated. Once you have located the constellation, you can position yourself accordingly to watch the display, with the meteors all appearing to fan out in all directions from the radiant.

Apart from the weather interfering there is also the phase of the Moon, which may cause issues. The Moon's reflective light can be an issue—the larger the phase, the greater the level of interference. This does not put off some observers, as bright meteors can still be witnessed. However, near to full Moon, fainter meteors are likely to be blotted out, and the observer may have to weigh up the position of the Moon and the radiant to see if it remains a viable option to observe on a particular night. That said, there is always the possibility that the Moon will set before the peak activity, which usually occurs after midnight, as the leading hemisphere of Earth turns into the debris field.

The Quadrantids

The first distinct annual shower in the northern hemisphere comes courtesy of the Quadrantids. This merits an explanation, as the shower is actually situated in a now defunct constellation. The name comes from Quadrans Muralis, a constellation created by the French astronomer Joseph Jerome de Lalande (1732–1807) in 1795. As a constellation, the name stood until 1922, before the IAU devised a definitive list of constellations, of which there are 88, with Quadrans Muralis not making it on to the final list. Quandrans Muralis was not alone, with other constellations also missing out. Prior to this, in 1908, Harvard had published the Revised Harvard Photometry Catalogue, which outlined its premise for 88 constellations.

The constellation of Quandrans Muralis was given its name to commemorate the wall-mounted quadrant mural that French astronomer Joseph Jerome de Lalande and his nephew, also an astronomer, Michel Lefrancois de Lalande (1766–1839), had used to measure star positions. However, this was not the only constellation not to make the final 88, with other formations also dispensed with. These included the rather grandiosely entitled Globus Aerostaticus, the Balloon; Machina Electrica, the Electrical Generator; Limax, the Slug; and Sciurus Volans, the Flying Squirrel.

Many other constellations also faced the axe at this time, including Cancer Minor, the Lesser Crab, which did seem to have more than just a small claim because of its larger counterpart Cancer Major, the Greater Crab. With so many constellations having major and minor counterparts—Leo Major and Minor, Ursa Major and Minor, Canis Major and Minor—sthis seemed plausible. Nevertheless, Cancer Minor was banished into obscurity, although the stars that make up the constellation, like many of the others lost star formations, still exist. In the case of Cancer Minor, it is a small grouping just to the left of what was subsequently deemed simply Cancer, the Crab.

The Quadrantids, though not surviving the cull, did survive in other astronomical aspects because of the shower that clearly emanates from within it, in an area that was encompassed by Bootes, the Herdsman, after Quandrans Muralis was no more. The radiant is situated near to the northwestern portion of the handle of Ursa Major, but, if the observer locates Arcturus, the bright star in Bootes, this will act as a good pointer. From here, the observer must simply trace the meteors back to the radiant point.

The Quadrantids shower has a very short peak to its activity, suggesting that the cluster of debris that Earth passes through early in January makes for a narrow band of material rather than a collection of fragments that are strewn out. The latter generates meteor showers that have a long, slow climb over the nights towards a peak of activity, then a general tailing off in numbers. The benefit of such a shower does afford the observer more time to catch activity, perhaps being able to be more selective of when to observe, given poor weather conditions or interference from the phase of the Moon.

The debris associated with the Quadrantids shower is most likely connected with the asteroid 2003 EH1, which, in turn, is thought to have a possible link to comet C/1490 Y1. In 1979, Isihiro Hasegawa pointed out that 500-year-old Chinese, Korean, and Japanese observations had recorded a bright comet in January of 1491 (C/1490 Y1) with an orbit similar to that of the Quadrantids. Astronomer Peter Jenniskens, a meteor expert with SETI, was able to calculate that the orbit of 2003 EH nicely coincided with the appearance of the shower. However, the link to comet C/1490 remains unproven.

With Bootes and Ursa Major clearly identified, the Quadrantids shower lasts for only a few days, January 1 to January 5, but your patience will be well rewarded with typically, after midnight, around 40 or so meteors per hour, known as the ZHR (zenith hourly rate). The meteors associated with the shower are generally bright, fast moving, with occasional reports of a bluish tinge. Entering the atmosphere at 40 km/s, the Quadrantids certainly make for a great deal of interest, with some of the meteors observed to leave a persistent dust train behind them, resembling a quickly evaporating con trail left by an aircraft. Just to light the enthusiasm further with this first major annual shower, the given rate of 40 meteors per hour as the ZHR is quite conservative, with the shower known to generate in excess of 100 per hour. It really just does depend on just how dense a region of the debris Earth passes through that night.

Crucially, although the shower lasts five days, peak activity occurs on January 4, but the observer should not just wait for the night of maximum activity but take every opportunity to observe, as—if you go on to actually submit reports to local or national astronomical societies—findings about the shower (for example, hourly rate, types of meteor seen) are used to formulate rolling data on the annual performance of a shower, and whether or not there seems, for example, to be a general decline in the amount of meteors seen, perhaps indicating a general weakening of the stream.

The Lyrids

The next most notable shower is the April Lyrids, associated with the constellation of Lyra, the Lyre or harp, with its main star Vega giving you an excellent pointer towards the radiant. This shower is associated with Comet Thatcher (C/1861 G1), discovered by Professor A.E. Thatcher on April 4, 1861. The comet's appearance in America during 1861 was regarded as an evil omen according to the *New York Herald*, a state of civil war having been declared that year.

No photographs have ever been taken of the comet due to its extensive orbital period of around 415 years, last visiting the inner

Solar System in 1861 when Thatcher first sighted the comet. Its next return is not expected until 2276. Debris left from the comet strikes Earth's atmosphere at 177,000 km/h.

Activity normally starts around April 16, continuing on until April 25, with the maximum of the Lyrids on April 21. On this date of peak activity, we can expect around 10–20 meteors at best per hour. Although not as attractive a spectacle as the Quadrantids, its springtime window offers potentially less cold weather and, as with any shower, there is always the possibility of a more lively display. On occasions, the shower has been known to produce tremendous displays of meteors, typically about every 60 years, with planets in our Solar System influencing the positioning and density of Comet Thatcher's dust trail through which Earth passes. In 1803, there are references in some records pointing to an outburst that year of 700 per hour. Other peak years of activity include 1922 and again in 1982, with a ZHR of around 90. Lyrid observational records go back 2600 years, and the debris from Comet Thatcher affords the Lyrids the title of being the only shower to boast the strongest annual shower of meteors from the debris of a long-period comet.

Lyrid meteors are generally swift and bright in nature, with luminous trains (trails). "Lyrid fireballs" have even been known to cast fleeting shadows because of their brightness, which on occasion have been said to surpass that of Venus. These fireballs have been seen to leave behind a smoky debris trail that can last a few minutes. The Lyrids have a counterpart, with a second installment emanating from the same constellation in June. The rates here are significantly lower, but the radiant does seem to be from same position as the April Lyrids. If the opportunity arises, this second shower is still worth watching, although, in recent years, the stream does look close to extinction as rates barely seem to make double figures.

The Eta and Delta Aquarids

The next shower is the Eta Aquarids, associated with the parent body of Halley's Comet. The Eta Aquarids span a number of weeks, making for a very protracted and elongated shower. Rates begin to rise around April 19, going on to span the remainder of the month and continuing deep into May, usually tapering to a finish by May 28. Peak activity for the shower is on May 4, although nights either side of this date may produce equal hourly rates after midnight, given the rise and descent from the peak, indicating the widespread nature of the debris associated with the Eta Aquarids.

This shower, through observational work during its span, would indicate that it is a declining one, with rates probably likely to be no more than 15 per hour at peak activity. This in itself poses a question as to why. With the parent body being a periodic comet, it would seem that the stream has a regular source of 'fueling' left by the ongoing returns of Halley's Comet. This may well be explained by a shift in the comet's orbit, with its second batch of debris encountered later in the year as the Orionids. This second encounter with the stream perhaps shows the more likely place in recent years for the comet to shed fragments. Also, probably through the influence of Jupiter, every 12 or so years there appears to be some interference with peak rates, with a noticeable drop in numbers.

The shower is also better seen from the southern hemisphere because it climbs higher in the night sky than in northern latitudes, affording the radiant to be not as obscured when positioned lower down on the horizon. Entering the atmosphere at 65 km/s, the Eta Aquarid meteors have been documented as yellow in color, some leaving long and persistent dust trains behind them. The radiant point in the constellation for the shower is the Water Urn.

Another shower springs forth from Aquarius during July, the Delta Aquarids. This shower is associated with Comet 96P/Machholz, discovered using a homemade cardboard telescope by American amateur astronomer Donald Machholz on May 12, 1986.

Activity for the Eta Aquarids starts around July 12 and lasts deep into August until around the 19th, with peak activity on July 27 although, again, spanning so many weeks indicates a plateau-like effect of meteors over the nights approaching and after that date rather than one distinct peak night of activity. Hourly rates around July 27 are slightly more favorable than the Delta Aquarids, possibly 20, and because the angle of the meteors hitting the atmosphere is different compared to Earth's passage through the debris back in May, the meteors virtually broadside Earth, striking the atmosphere at 40 km/s.

The Perseids

August contains one of richest and most consistent meteor shower performers, the Perseids. The Perseids are a reliable presence and nicely situated in the warmer months, making observing quite pleasurable. That said, there is much to be said about the clarity of the sky during winter, with the clear crispness that this offers, lost in the summer months, with a haze to the late evening and night sky often left lingering from the warm daytime (Fig. 9.1).

Commencing on July 17 and running until August 24, with peak activity night on August 11/12, the Perseids go somewhat against the trend, delivering a noticeable peak compared to other showers that have a lengthy run of days but with a plateau of activity. Situated in the constellation of Perseus, the shower is

Fig. 9.1 Astronaut Ron Garan took this image of a Perseid meteor from the International Space Station, August 2011. Look for a small streak of light just right of center (courtesy of NASA)

associated with Comet Swift-Tuttle, which deposits debris every 130 years or so during its orbit around the Sun. Unlike other meteor showers, a distinct cloud of debris generates the Perseids, and, as long as the comet remains in existence, this cloud will be continually seeded. Comet Swift-Tuttle has an interesting history, with a strange, almost oblong-shaped orbit, which takes the comet outside the orbit of dwarf planet Pluto at its greatest distance from the Sun, swinging back around and inside the orbit of Earth when it is closest to the Sun.

The comet was discovered by American astronomer Lewis A. Swift (1820–1913) on July 16, 1862, during the height of the Civil War. At the time, it was also sighted several days later by American astronomer and Union naval officer Horace Parnell Tuttle (1837–1923). Tuttle continued to make observations of the comet throughout the war, with the title 109P/Swift-Tuttle subsequently tagged to it as a joint find.

However, it was Italian astronomer Giovanni Schiaparelli (1835–1910) who noted that the orbit of the comet could well be linked to dust showers seen from Earth every August. Further to this, and following on from studies conducted by American amateur astronomer Gary W. Kronk, British astronomer Brian G. Marsden (1937–2010) confirmed Kronk's connection between Comet Swift-Tuttle and a comet that was discovered by Jesuit missionary Ignatius Kegler on July 3, 1737. Presumption followed that Comet Swift-Tuttle had indeed been orbiting the Sun for centuries.

The year 1982 saw eager anticipation for the return of Comet Swift-Tuttle, but it was to be a spectacular no-show. A proportion of astronomers feared that the bulk of Comet Swift-Tuttle's mass had evaporated during its last encounter with the heat of the Sun and that subsequently the comet had possibly disintegrated. However, 10 years after that disappointment, the comet showed up, being first sighted and confirmed by Japanese amateur astronomer Tsuruhiko Kiuchi, who rediscovered the returning body using high-powered binoculars on September 27, 1992. It was on this return that evidence was established that the comet Kegler had sighted back in 1737 was Comet Swift-Tuttle, with its closest approach to Earth being 17 days later than predicted by astronomers. However, the return of the comet did present astronomers with a much darker and potentially more serious problem. Given the orbital data now to hand, it seemed possible that, at a later point in time, Comet Swift-Tuttle could well be on a collision course with Earth.

If this was the correct choice, then with the Perseid shower of 1981 or, indeed, showers either side of 1981, the ZHR could be expected to significantly climb, with the stream freshly rejuvenated with new cometary material. The years of 1976 to 1983 showed an increase in shower activity, with rates after 1983 dropping back from the heightened ZHR of 90 to the more regular ZHR of 65. At its peak, the ZHR had reached 187 during that 1976 to 1983 span.

However, during the 1981 predicted return there was no sign of Comet Swift-Tuttle and, despite great enthusiasm by meteor observers in the astronomical community, this was not shared by comet enthusiasts, and following the comet's no-show questions duly followed.

Although there had been a distinct and noticeable rise in meteor shower activity, as Marsden's calculations predicted, where was the comet? Marsden revised his predictions, switching from the original choice of 1750, which he had based his calculations on, and changing it to the comet sighted in 1737 instead. With new calculations based on a different comet, 1992 was earmarked as the year that would see Comet Swift-Tuttle's return to perihelion. During the summer of 1992, the comet was duly picked up in the night sky and, although this wasn't the best appearance that the comet had presented us with, it bore out Marsden's revised calculations.

For meteor observers, on hearing the confirmation of Marsden's predictions, an air of excitement surrounded the likelihood that the Perseids shower would reflect the return by generating a spectacular outburst. With observers across the globe waiting in eager anticipation for the nights of peak activity, the scene was set. The return of Comet Swift-Tuttle did not disappoint, with all expectations not only met but surpassed beyond observers' wildest dreams, as meteors filled the night skies during August 1993. The ZHR rocketed, with 200–500 meteors per hour readily observed, making 1993s Perseids shower one of the most watched and documented in history.

Further research has determined multiple radiant points within Perseus. The early work on these multiple radiant points within Perseus was attributed to British amateur astronomer William Frederick Denning (1848–1931) who, in 1879, identified different radiant points for the Perseids. Denning discovered that two simultaneous showers were taking place, radiating from near to Chi and Gama Persei in Perseus. Subsequent studies, most notably a study conducted between 1969 and 1971 by observers in the Crimean Peninsula, confirmed Denning's conclusions.

Whereas showers tend to produce a similar vein of meteors, the Perseids can also produce a great variance in meteor types. In 1953, observations made by A. Hruska (Czechoslovakia) discovered that, on viewing the Perseids over a period of nights, the batch on August 8 to August 11/12 were brighter than those on August 12/13, and then noticeably fainter by August 14/15. Three years later, Z. Cephecha (Czechoslovakia) discovered meteors to be brighter on August 6/7 compared to those of August 13/14. Further studies in the 1980s and again in the 1990s came to similar conclusions.

These variances could only mean that the filaments deposited by Swift-Tuttle on its various returns, plus the actual amount of fragments left behind, were massively inconsistent. This would account for the different types of meteors being recorded, and also the ZHR. A further study by M. Plavec (Czechoslovakia) analyzed a staggering 8,028 Perseid sightings between 1933 and 1947. Plavec noted that 45% of Perseids left distinct and persistent trains in 1933, with this value increasing to 60% in 1936 but decreasing to 35% in 1945, with a rise to 53.3% recorded in 1947. Despite the findings, no further explanation was offered, aside from the different chemical compositions within the debris left by the comet and the actual amounts it was leaving.

Reports of the actual Perseid shower itself date back to records kept by the Chinese in A.D. 36 during the time of the Eastern Han dynasty, which ruled from A.D. 25 until A.D. 220. During this time in Chinese astronomy, great progress was made, including what is believed to be the earliest manufacture of a seismograph by mathematician and astronomer Zhang Heng (A.D. 78–139).

The Romans linked the meteor shower with their fertility god, Priapus, stating that the shower was an ejaculation by the god to seed not just the Roman landscape but its livestock, with the hope of fruitful harvests and ripened fruit from the orchards in the seasons to come. The Romans saw the annual outburst from Perseus as a baptism of the land, with a mixture of water, honey and wine, ensuring essential nurturing of pastures. In Catholicism, the Perseid meteors were deemed to be the 'tears of Saint Lawrence,' where every year on August 10, to mark the saint's martyrdom in A.D. 258, tears flow. This occurrence also aligns with the Mediterranean folk legend where St. Lawrence, having been burned alive, sees the instrument of his death, a gridiron, eject sparks into the night sky on August 9/10, with the cooling embers falling to the ground, to be found among the flora and fauna. These glowing embers as they further cooled on the ground were known as the "coal of St. Lawrence".

In 1835, Belgian mathematician and astronomer Adolphe Quetelet (1796–1874) documented the shower as having a radiant from Perseus. Quetelet, founder and first director of the Brussels Observatory, published the first catalog of meteors and is credited as independent co-discoverer of both the Orionids and the Quadrantids in 1839. In 1862, Italian astronomer Giovanni Schiaparelli established the link between meteor showers and comets, proving the association with cometary orbits and the appearance of annual showers.

Comet Swift-Tuttle certainly generated plenty of questions with regard to its orbit and depositing of debris, but the most intriguing and most alarming was the question posed that, given the comet's orbit, could it eventually collide with Earth? British astronomer Brian G. Marsden initially expressed a possibility of such a collision taking place. However, he subsequently retracted the claim, concluding that there was no danger of either Earth or its closest neighbor the Moon being struck.

In August 3044, Comet Swift-Tuttle is due to make a close passing of Earth, less than 1.6 million kilometers. Traveling at 61 km/s, were Comet Swift-Tuttle's orbit to have been altered by planetary influences, or a glancing blow from another body, it would plow into Earth with results akin to extinction level. The energy release calculated from the collision has been estimated at the equivalent of 27 times that of the object that wiped out the dinosaurs!

The Draconids

The Draconids shower during October generates around 10–20 meteors per hour. With its radiant in Draco, the Dragon, rates increase from around October 6, with activity noticeable until October 10. This short shower has its peak ZHR on October 8. As with other showers with a relatively low rate during the night of maximum activity, recording of the shower remains of equal importance against more bountiful showers, in order to document any outbursts or decline. Meteors associated with the shower have been generally noted as slow-moving and yellowish in appearance, striking Earth's atmosphere at 20 km/s.

The Draconids, also unofficially known as the Giacobinids, are associated with Comet 21P/Giacobini-Zinner, with its radiant point in the head of the Dragon. Locating the head for the uninitiated will take a little doing, as Draco spreads itself nearly halfway around the north celestial pole, the constellation being the eighth largest of the 88.

Fortunately, the head of the Dragon contains two bright stars, Eltanin and Rastaban, making up the eyes of the beast. Despite its meager peak ZHR of 10 to 20, the shower has been known to generate a substantial amount more, with showers in 1933 and 1946 producing literally thousands per hour, as Earth passed through a very dense part of the debris stream associated with the shower. Further outbursts have also been documented, notably in 1998, 2005, 2011, and 2012.

Comet 21P/Giacobini-Zinner orbits the Sun every 6.5 years, with its path periodically straying near to Jupiter. A substantial stream of particles was believed to be ejected from the comet in 1900, and, despite its intermittently close association with Jupiter, the planet does not seem to have a great effect on the debris, with the stream structure remaining relatively intact.

The Orionids

The Orionids shower sees activity increasing from October 16 onwards, with meteors sighted until October 26, and the night of maximum output occurring on October 20. With the sky after midnight beginning to show the winter constellation, the unmistakable figure Orion, the Hunter, gradually looms into view, providing the watchers of this shower with the ZHR at a maximum of 20–25. Known to produce yellow and green meteors, these fast-moving shooting stars travel at over 66 km/s, but rates have never been exceptional and, if anything, some years have been thoroughly disappointing. However, equally, some years have been bountiful.

The discovery of the Orionids dates back to 1839 and is credited to American E.C. Herrick, who made a statement about activity in Orion for the period October 8 to October 15. In 1840, he added more weight by making a further statement to complement the first. However, it was British astronomer Alexander Stewart Herschel (1836–1907) who made the first documented observation of the shower on October 18, 1864. Herschel, grandson of William and son of John and Caroline, recorded 14 meteors appearing that night that radiated from near the constellation of Orion. Hershel, who conducted pioneering work in the field of meteor spectroscopy, confirmed his findings a year later, on October 20, 1865.

Some astronomical finds provoke great debate, and one such argument ensued over the actual location of the Orionids radiant, which we now know are centered to the northwest of Orion's top left red giant star, Betelgeuse. Betelgeuse is easily located, being very prominent in the night sky.

William F. Denning and American astronomer Charles Pollard Olivier (1884–1975) disagreed over the Orionids' radiant. Denning claimed that the radiant point moved one day to the next, so it was not a constant site for the meteors. Olivier challenged the claim, stating that the radiant was from a fixed point. After much debate, new technology prevailed, finally bringing the argument to a conclusion, favoring Olivier.

The Taurids

November boasts two showers of note, the first being the Taurids, in the constellation of Taurus, the Bull. This shower is associated with debris left by Comet Encke, which orbits the Sun every 3.3 years and has been observed during all of its returns since 1818 with the exception of one, in 1944, during World War II. Encke, first observed by French astronomer and prolific comet hunter Pierre Mechain on January 17, 1786, is also thought to be one of many other fragments left over from a much larger body. There have been suggestions of a link between the Tunguska event and Comet Encke, most notably the proposed link made by Slovak astronomer L'ubor Kresak (1927–1994) in 1978. The discoverer of two comets himself, 41P/Tuttle-Giacobini-Kresak and C/1954 M2 (Kresak-Peltier), Kresak believed the body responsible for the Tunguska event was a fragment of Comet Encke.

The debris left by Comet Encke has been spread out over a fairly wide area through which Earth passes, with the stream's components comprising much weightier objects than other streams. These larger objects, pebble-sized, can make for, on occasions, decent fireball activity, affording the Taurids another title, the shower that produces Halloween fireballs. With the Taurids generally underway a full fortnight or so before Halloween, a prediction was made in 1993 that 2005 would be a particularly good year for Halloween fireballs, and it was, with many being sighted. The prediction was made by astronomer David Asher at the Armagh Planetarium in Northern Ireland, stating that, every so often, Earth passes through a swarm of slightly larger debris within the stream, generating these very bright meteors. Similar increased rates of Halloween fireballs were observed in 1998. On occasions, even larger objects have been noted, with bolide activity also linked with the Taurids, documented as being as bright as the Moon and leaving smoke trails.

Due to the gravitational effects of several planets in the Solar System, most noticeable Jupiter, the Taurids has become an elongated shower, with two separate streams having been identified, those of the Northern Taurids and the Southern Taurids. The Southern Taurids start earlier than the former, around September

10, with activity until around November 20. The Northern Taurids are active from October 20 until around December 10.

The radiant of the Northern Taurids is close to the Pleiades star cluster, while the Southern Taurids' radiant is close to the stars Omicron and Xi Tauri. The hourly rates at maximum are not exceptional, with this now very old stream managing a ZHR of 5–10. The bulk of meteors observed from the Taurids' showers are white in color with a lesser percentage yellow. Other colors, including blue, green, and orange, have been documented.

Apart from speculation over the link to the Tunguska event, this shower also has another interesting claim. In November 2005, American astronomers Dr. Robert Suggs and William B. Cooke observed an impact on the Moon's surface from a Taurid meteorite. While testing out a 10-inch telescope fitted with video recording equipment, specifically designed to monitor lunar surface strikes by meteors, a meteorite impacted. After subsequent analysis of the strike by Suggs and Cook, a link was established to a meteorite heralding from the Taurid stream, making this video recording the first ever of its type.

The Leonids

Of all the annual showers that could deliver a sky full of shooting stars, it is the Leonid shower in November that probably holds the most in its armory to surprise. With activity in the Leonids commencing around November 14 and concluding a week later on November 21, it is this shower that is so important in history and can deliver so many dazzling displays of cosmic fireworks. However, with the shower peaking around November 17, the Leonids by the same token can be equally disappointing.

The shower is held in great esteem within the astronomical community, for the discovery and ensuing study of the stream marked the start of meteor astronomy. Work on the Leonids laid down many of the fundamental principles for the scientific study of meteors, from the basic mechanics to fluctuating stream activity and shower evolution. This revolution in thinking also dispensed once and for all with the arguments that the phenomenon was merely one of an atmospheric nature, and nothing whatsoever to do with space.

The first accounts of the Leonids shower dates back to A.D. 902. In A.D. 898, the yet undiscovered comet that was to become the source of the debris for the stream crossed the path of Earth's orbit for the first time. Four years later Chinese astronomers and observers in Egypt and Italy reported the first Leonid storm, an outburst that was to continue to be intermittently recorded in the centuries to follow where, on one occasion, sky watchers proclaimed that the "stars fell like rain".

In 1630, several days after the death of Johannes Kepler on November 15, 1630, the sky lit up with shooting stars from the then unknown Leonids shower, at the time reckoned by some observers to be a salute from God to the German astronomer.

In 1799, German scientists and explorers Alexander von Humboldt (1769–1859) and companion Aimé Bonpland (1773–1858) observed an outburst from the Leonid shower in Cumana (latter-day Venezuela). Humboldt and Bonpland documented what they had seen in detail, releasing their findings to the scientific community. Probably quite unwittingly, they had recorded an event that was to become quite infamous in the world of meteor studies. There is speculation that previous to this outburst from the shower in 1799 there had also been quite a display in 1766.

However, it wasn't until the night of November 12/13, 1833, and with particular attention to the early morning hours of November 13, that the skies literally filled with meteors, with some reports at the time claiming that Judgement Day itself had arrived.

With both the scientific and public imagination fueled by the Leonids outburst of 1833, the November skies were duly observed with great intent and, despite more brilliant displays, some years were to prove incredibly frustrating, with the shower not living up to expectations. Many theories were put forward as to the erratic nature of the shower, including the highly imaginative explanation of electrified air interacting with the onset of falling temperatures and the subsequent coldness of the mornings, courtesy of the U.S. telegraph of Washington, D.C.

However, it was American astronomer Denison Olmsted (1791–1859) who gave the most credible explanation. Olmsted, using much of his own personal observations, presented his findings to the *American Journal of Science and Arts*, with his work published early in 1834 and again in 1836. His findings were to establish precedence for the future of meteor observation; indeed, his work became internationally recognized and, to this day, many credit Olmsted as being the founder of meteor science. His observations shifted meteors away from the then current thinking that their presence was Earthly in nature, proposing space as their origin.

Olmsted had noted that the Leonid outburst of 1833 was of short duration, and that it had not been witnessed from Europe. Olmsted also noticed that there was a central point in the sky from which the meteors came, a radiant that appeared to be in the constellation of Leo. He concluded, by making a comparison study to the November shower of 1832, where abnormal rates had been seen across Europe and the Middle East, that the meteors were in fact the result of Earth passing through a cloud of particles in space and had nothing whatsoever to do with atmospheric disturbances. Although Olmsted did not explain the exact nature of the particles, it did pave the wave for further studies into the potential gathering of 'cloud-like' formations of debris in space that, at regular intervals throughout the year, Earth's orbit would pass through.

The Leonids were to put on another fine display in 1866, with observers documenting hourly rates as high as 5000, with a reduced rate of around 1000 in the following year. A similar ZHR of 1000 was to be generated in 1868.

However, the question still remained as to the origin of the Leonids and, if it were a cloud of debris (which seemed to be the now widely accepted belief), where did all this material come from? This puzzle began to unravel itself on December 19, 1865, when German astronomer E.W.L. Tempel made the discovery of a new comet near Ursa Major. The comet was subsequently sighted and documented in the following year, on January 6, 1866, by French astronomer Horace Parnell Tuttle. The comet was duly named 55/P Tempel-Tuttle. This meant that Tuttle had now discovered two comets linked with two showers, the other being 109P/Swift-Tuttle linked with the Perseids.

After its closest approach to the Sun on January 12, 1866, the comet faded and disappeared but, using the information gathered from this particular encounter, calculations were made concluding that 55/P Tempel-Tuttle was a short period comet. A link was subsequently established between heightened activity in the Leonids shower and to that of the orbit of the comet. Further to this work, in 1867, Austrian astronomer Theodor von Oppolzer (1841–1886) calculated an orbital period for the comet of 33 years, meaning the next display of any note should occur before the turn of the century, in 1899.

French mathematician Urbain Le Verrier (1811–1877), whose specialized area was celestial mechanics, subsequently computed an accurate orbit for the appearance of the Leonids stream, with von Oppolzer, Schiaparelli, and the director of the Königsberg Observatory, Dr. Carl Friedrich Wilhelm Peters (1844–1894) all independently reaching similar conclusions.

All attention now turned to the comet's return in 1899, and to whether all the research and conclusions would be borne out with another spectacular outburst from the Leonids. However, in equally spectacular fashion, the shower failed to perform. Founder of the American Meteor Society Charles Pollard Olivier stated that the non-event was "the worst blow ever suffered by astronomy in the eyes of the public." With the shower failing to deliver, explanations were immediately sought. The general consensus of opinion was that the cloud of debris left by the comet had obviously shifted its position, out of the orbit of Earth. Further conclusions were that previous outbursts from the shower must have been merely freak passages of Earth through cometary filaments, and that showers of this nature were now confined to history, with no outbursts ever likely again.

Despite an initial good showing of the meteors in 1898, all had seemed well with the stream, so questions as to the 1899 show rumbled on. There was some cheer to be had despite the generally poor showing of 1899, with 1901 reports from the western half of the United States reporting the ZHR at 300–400 per hour.

However, further studies did later reveal the seemingly elusive answer. It transpired that the stream had shifted its position, influenced by near encounters, first with Saturn in 1870 and then Jupiter in 1898, throwing due date interactions of any substance with Earth out of kilter. Attention based on these new findings was now redirected to 1932, with the next return of Comet

Tempel-Tuttle. Heightened rates during the return of the comet were observed, in the region of a ZHR of 200. However, while this provided short and elevated hope of another outburst, it eventually only served to confirm how erratic predictions of the stream's interaction with Earth had become. The year 1932 was also to mark a decline in the Leonid rates and, despite some fairly lively displays up until the end of 1939 (30–40 per hour), the years now appeared to show that the shower had reached a plateau phase in terms of rates.

With curiosity still as heightened as with previous returns, attention focused on the next return of Comet Tempel-Tuttle in 1966. Previous returns had damaged expectations for a burst of meteor activity, and, once again, the shower failed to deliver. At least, though, there was now a reluctant acceptance of what was turning out to be a regular non-event. The year 1966 could only manage a near parallel to 1899, with around 100 meteors per hour being fairly widely observed. Despite the generally poor showing, and in keeping with the now established erratic nature of the Leonids, once again, the western half of United States was treated to a morning display of meteors not to forget. In the pre-dawn skies of November 17, observers were treated to the rare spectacle of witnessing what was later estimated to be 40–50 meteors per second. Rates remained elevated for a time over the coming years, with a dip in 1970 and then a further upturn in rates in 1971.

The uncertainty of the shower took a major turn in 1981 thanks, most notably, to the work of research scientist Don Yeomans, working at NASA's Jet Propulsion Laboratory. Yeomans embarked on an in-depth study between the relationship of the Leonids and Comet Tempel-Tuttle. With extensive study into the distribution of debris left by the comet, Yeomans determined that there was a lag from the ejected cometary material that fanned out to a much wider path than the comet's orbit. Factoring in the influence from the solar wind, and the previously discovered influence from Jupiter and Saturn, Yeomans concluded that anomalies in expected ZHR were inevitable, further concluding that roughly 2500 days near to or soon after perihelion of Comet Tempel-Tuttle, and at a given distance that had to fall within certain parameters, then and only then would it appear that conditions were at their optimum to produce a shower of any significant intensity, an outburst. However, with an inconsistency in the cometary particle distribution, there would always remain a great deal of uncertainty about the outcome.

The 1990s produced more erratic displays from the Leonids shower, with the now expected normal ZHR of around 10–15 an hour only reasonably heightened during 1994 and again in 1995, with 40 meteors per hour recorded. As with other years, the shower didn't fail to deliver a special showing for some parts of the world. In 1997, observers watching from the Canary Islands were treated to a ZHR upwards of 2000.

In 1999, Robert H. McNaught (Research School of Astronomy and Astrophysics, Australian National University) and David Asher (Armagh Observatory) published a paper confirming the connection between Comet Tempel-Tuttle and the Leonids. The paper noted that Jupiter would slightly pull the comet into a brand-new orbital track, catching the comet in its gravitational fields either on approach to the Sun or on its outward bound journey away from it. In doing so, the particles deposited from the comet's tail on each separate return would be left within the already established debris cloud as a series of ringlets or filaments. Over a period of time some of these ringlets would subsequently diffuse and merge with other ringlet formations and, when added to by subsequent new debris, place substantial deposits in the orbital path of Earth.

On the basis of this analysis, McNaught and Asher predicted that during 1999, because of a certain filament left by the passage of the comet in 1899, rates would climb to 1000 meteors per hour. As with previous years, it was again a case of playing the waiting game to see what transpired. Any doubt was soon dispensed with and, to the collective relief of the scientific community, McNaught and Asher were proved correct, with the only minor flaw being the underestimated ZHR, which surpassed 3000.

In 2006, an outburst was predicted from the 1932 trail of debris left by the comet, with rates higher but not overly increased, being a ZHR of 70–80. In 2007, the ZHR was higher than average, but halved against 2006. The following year saw a rise in ZHR to over 100, this particular increase associated with Earth passing through a filament left by the 1466 return. 2009 gave an impressive display to some observers across the globe, with a ZHR in excess of

500. After this heightened run of meteors from the Leonids stream, the shower went into decline again, but history dictates that, given its varied record, the Leonids are far from being inactive, and it is surely only a matter of time before Earth passes through another filament.

On a good, clear, moonless night, we can still expect the Leonids to produce a ZHR of up to 20 during peak activity. The shower is generally characterized by fast meteors, mostly showing a distinctive blue or green color in appearance, with many leaving persistent dust trains. With the meteors traveling at 70 km/s, the shower radiant lies close to the star Algieba in Leo—not the brightest star in the constellation, but nevertheless relatively easy to spot under good skies. The bottom line with the Leonids is that it is always worth the effort to observe... just in case!

The Andromedids

November meteor showers wouldn't be complete without the inclusion of the Andromedids, to which a quite extraordinary tale is attached.

The Andromedid shower is a very minor affair, which is often overlooked as, sadly, other such minor showers are. Over the last century, observers of the shower have struggled to document any activity of note, with sporadic meteors perhaps being inadvertently linked with the shower or, by the same token, also not logged as being part of the shower. However, history makes for a need to keep monitoring the stream if at all possible. The reason for the Andromedids' demise is that Earth no longer intersects the denser region of the debris stream. We know the stream to still be in existence, and in the distant future the debris will cross our orbit. In the interim period, less than 1 % of all meteors now observed on the night of maximum, in late November, are Andromedids.

The Andromedids are linked to the legendary Comet Biela, a comet that when originally sighted kept astronomers guessing, because it had been logged so many times as a new comet, only to be confirmed much later on that all such sightings were, in fact, of one comet, Biela. Biela was discovered by a French amateur astronomer Jacques Leibax Montaigne (1761–1745) on March 8, 1772. Montaigne was only able observe the comet for around a month before it was lost. Because of the shortness of its appearance, Montaigne, who also discovered two other comets in his time, was not able to establish whether this comet was a rogue or whether it actually held a stable orbit around the Sun.

Rediscovered by French astronomer Jean-Louis Pons on November 10, 1805, this time the comet was sufficiently on view for Pons to make a more detailed study of it, establishing that first and foremost the comet was not a rogue and, being all but absolutely certain, calculated an orbital time frame that would give Biela a return schedule. Any doubt was dispelled on February 27, 1826, when Wilhelm von Biela sighted the comet in the constellation of Pisces.

Despite the comet's visits not being recorded in the years between 1805 and 1826, an orbital period of 6.6 years was assumed, and, in 1832 as forecast, it appeared again. Although the comet was not picked up again until 1845, it was still putting in appearances in line with the suggested orbital period. With Montaigne's initial discovery and observations overlooked, credit for the find went to Biela, and the comet was duly named Comet Biela.

During Comet Biela's next return in 1845/46, it became evident that something unusual was taking place with it. On January 13, 1846, American astronomer Matthew Fontaine Maury noted that the comet now had two separate components to it, as if Biela had split in two. These two components also seemed to be gradually becoming more distant, with a distinct gap now developing and becoming increasingly larger as spring 1846 approached. Speculation about what was to become of the comet became rife in the scientific community, but as it subsequently faded into the darkness from sight, both fragments were still visible.

Right on schedule, Biela's Comet returned in 1852, and, just as the components had disappeared from view in 1846, both re-emerged from the darkness. However, this return was to be the last time that Biela's Comet was sighted. Poor observing conditions in 1859 hampered any attempt to recover the comet so there was much uncertainty as to whether or not it did actually return.

Astronomers eagerly awaited better skies and the comet's return in 1866, but, despite much scouring of the skies, there was no sign of Biela, and it was deemed lost.

There was much speculation in the years to follow that other comets that were sighted may well have been either of the components of Biela, but nothing was ever confirmed. The comet's fate is most likely to remain a puzzle, with perhaps a collision, total disintegration or a shift in orbit responsible for its demise. Still, for many scheduled returns later, there was always some hope that it would reappear, but it never did.

Despite being a relatively minor annual shower in recent times, the Andromedids was first documented on the night of December 6, 1741, when, across the skies of St. Petersburg in Russia, an outburst from the shower was observed. Subsequent years spawned further displays of note, but nothing in comparison to the display the Andromedids produced in 1872.

Austrian astronomer Edmund Weiss made the prediction that the Andromedids would make more than just an appearance in November 1872, with the dates 27 and 28 marked for major activity. Weiss' confident and rather precise predictions came true, with a spectacular storm generated late in the month. Hourly rate observations from the year suggest that something in region of 7000 per hour had been generated. Another storm of near equal activity was observed on November 27, 1885. With this great surge in hourly rates, there was much speculation that perhaps Biela's Comet had indeed disintegrated, and what observers were witnessing from these outbursts was Earth passing through the remnants of one if not both components.

However, Peter Jenniskens' book *Meteor Showers and their Parent Comets* threw a different light on the storm. Jenniskens suggested that the storms of 1872 and 1885 were merely part of Biela's cometary dust debris left by the comet as it rounded the Sun on previous visits, those made in 1846 and 1852. Jenniskens further noted that the separating of Biela into two segments hadn't released a great amount of debris. Jenniskens concluded that, on this basis, the Andromedids would still have stormed even the comet hadn't split. It is perhaps fair to comment that the outbursts were possibly the effect of Earth merely passing through a sequence of particularly dense clusters of cometary debris. This would also account for years when Earth had passed through lesser and, in some cases, virtually no debris.

The Geminids

The last and most reliable annual shower is the Geminids, although the Ursids shower slightly later in December can produce fair hourly rates at maximum.

The Geminids shower sees activity commence early in December, on or around the 7th, lasting until December 17, with the peak night for activity being December 14. The Geminids and the Quadrantids are the only two major annual meteor showers to be associated with asteroids rather than comets, with the Geminids linked with the rather strange world of 3200 Phaethon, which is thought to be a Palladian asteroid, although some have suggested that the asteroid is an Apollo.

With an orbital period of 524 days, 3200 Phaethon is known in some circles as the "rock comet" and since its discovery has been the source of much debate. First, 3200 Phaethon displays all the characteristics of a comet in nature, with observations made by NASA's STEREO spacecraft showing dust tails from the object, possibly caused by fractures on the surface, these cracks appearing as the Sun heats the body as it nears perihelion. Secondly, its apparent orbit is more in line with that of a comet than an asteroid.

Measuring just under 5 km wide, 3200 Phaethon has a very close orbit around the Sun, with the object literally baking and crumbling in the heat as it passes, causing it to lose some of its mass in rubble during subsequent passes, spreading this debris in its orbital path. December sees Earth pass through a fair swathe of these fragments, generating the Geminids shower. Striking our atmosphere at 35 km/s, the shower, first discovered in 1862, is thought to be intensifying in nature as the years pass, with maximum activity now capable of producing in excess of 100 per hour.

Discovered by the Infrared Astronomical Satellite (IRAS) , its provisional tag was 1983 TB, later revised with its numerical number and name in 1985. Working on data sent back from IRAS, astronomers Simon F. Green and John K. Davies located 3200 Phaethon on images taken by the satellite on October 11, 1983, announcing their discovery on October 14. Optical confirmation followed courtesy of Charles T. Kowal (1940–2011), who reported the find to be an asteroid in nature.

3200 Phaethon, though, remains a puzzle—an asteroid for sure, but with a comet's behavior. With classification a necessary tool in astronomy in order to categorize and create order, 3200 Phaethon's own classification presents a dilemma. Fortunately, or unfortunately, 3200 Phaethon is not alone, as other finds have subsequently created many subsections with revisions having been made to accommodate new finds that simply don't fit into any category. However, despite the cometary tag, 3200 Phaethon has been classified as a B-type asteroid, because of its dark material. Since other objects of this nature have been found, perhaps the debate will cease as to exactly where 3200 Phaethon sits in the overall classification of cosmic debris. Obviously not a traditional comet, 3200 Phaethon could be summed up as a split-asteroid, neither one thing or the other, but whatever we make of it, it just adds to the rich and diverse tapestry of all known cosmic debris.

As for the Geminids, its apparent newfound intensification now places the shower as one of the year's must-see astronomical events with slow, easy to spot meteors making for a glorious display under the crisp, dark, winter skies. Producing a multi-colored display, the ratios give the majority of meteors as white, with over a quarter displaying yellow, and around 10% blue, red, or green in color. At their moderate speed, which is generally a lot slower than most meteors associated with annual showers, these are easy to spot, and can be traced back to the radiant, which is very close to the bright stars Castor and Pollux. The shower has often been known to produce bright meteors, and even fireballs.

The Ursids round off the annual displays of most notable showers, commencing on December 17 and running until December 26, with the peak on December 22. Discovered by William F. Denning around the turn of the twentieth century, the Ursids have a rather checkered history, completely overlooked at times, then of seemingly great interest, especially when there are short-lived outbursts, raising its normal rates from 10 to 15 per hour at maximum to over 100. The shower is associated with

Comet 8P/Tuttle, a periodic body rounding the Sun every six to seven years. However, influence on the comet from Jupiter, plus a series of other factors, have meant that, in recent times, outbursts have become rare indeed as the stream is drawn away from Earth, so our orbit no longer intersects with many of the Ursids' debris fragments.

The radiant is located within the constellation of Ursa Minor and, under the clear, crisp winter skies, can still make for a fair display.

How to Record Sightings

As we have established, the actual observation of meteors requires very little preparation, although there are some things to consider before settling down to watch the skies. Having sought out a good vantage point with not too much in the way of interference in artificial lighting and a view that affords the observer as much of the sky as possible with the best panoramic view, a few other pointers are in order to make the most of the hours ahead.

Some observers may find standing for a long time difficult, so being seated would be helpful. A garden chair would suffice. Others prefer to lie down on a lounger although, on warm summer nights, the temptation to rest one's eyes may result in falling asleep, with little observational work then following. Whatever is chosen, the observer must feel comfortable. Feeling uncomfortable will often result in any observations being cut short, so, as basic as this may seem, do be happy with whatever position is assumed! Naturally, a chair comes into its own if the observer is watching a shower, as it's only a case of positioning the seating in the direction of the radiant, sitting back and looking up. If observing random meteors when a shower is not expected, it may be prudent to stand, as this will allow the observer to scan all the sky without having to constantly get up and move a chair around to a different position.

Although it may seem logical to have some sort of refreshments, continual cups of hot liquid will in turn mean frequent trips to the lavatory. Having allowed your eyes to be accustomed to the darkness of the night (dark adaptation), leaving a settled position to go into a lit house will ruin this, not to mention the cooling down of hot liquids in the observer's system, which will make for a colder than normal feeling as the outside temperatures decline as the night passes.

Dark adaptation is important as, over a period of half an hour, the observer will begin to notice much smaller and fainter objects in the night sky that, before dark adaption, just simply weren't visible. The brain has now switched on to this low level of light, affording not just a better visual intake but also heightening the sense of hearing, too, as the observer almost integrates with the surroundings (Fig. 9.2).

Whether using pencil and paper or an electronic voice recorder, there are a number of points that must be recorded to make your observation of meteors valid and of use to those collecting the data. It may also be prudent to cover any flashlight being used with perhaps a piece of red cellophane; this will allow the observer to record sightings without losing dark adaption, which would surely be lost by the brightness of the flashlight's white beam.

It is quite likely that astronomical societies will provide their own log sheets for observing meteors, but, if not available, here are

Fig. 9.2 A star field in the constellation of Cepheus. Allow your eyes to grow accustomed to the night sky (courtesy of NASA)

some guidelines to help record any sighting. Also remember: proper log sheets or not, the information record, if done correctly, is just as valuable.

Here, now, are some additional recording tips:

Date

Note when the watch commences and when the watch ends. Record the data as follows. If you commence observing on the evening of August 11, 2014, and finish observing in the early hours of August 12, 2014, this should be written down as 2014/08/11–12.

Start Time/End Time

Record the start of your watch and the end of watch in Universal Time (UT). Remember to take into account any time zones that may apply.

Your Name

Self-explanatory!

Location

If possible, exact coordinates would be useful, and many modern cell phones are able to pinpoint exact locations using GPS, for example. If this isn't available, try the Internet or indeed an ordinance map, which should provide longitude, latitude, and approximate height above sea level for the location.

Sky Conditions

The observer will need to determine what sort of sky conditions the observations were made under. It could well be that the observing sessions started with clear skies, but cloud builds during the night to partially obscure the view. By the same token, skies may start somewhat cloudy and then clear later on as the night progresses. Also, the Moon may cause some interference.

It is always advisable to try and select nights where there is no full Moon, as the reflected light from the Moon will virtually ruin the observing conditions, not allowing the observer to see fainter meteors. It is understandable that with a potentially unsettled period of weather, and having to watch for the appearance of a full Moon, this does limit the time available to actually go out and observe, which is why every opportunity to do so should be taken. Over the months and hopefully years ahead, observers who have just started to watch the skies will become very skilled at judging these affairs, so please be patient. Record your sky conditions at regular intervals, estimating cloud cover in percentages, with the onset of any mist or fog.

NELM

Apart from conditions, you will need to record the naked-eye limiting magnitude (NELM) . This is the clarity of the sky being observed, not the actual conditions. The NELM is defined as the magnitude of the faintest star that can be seen using averted vision (where you're looking not directly at a chosen object like a star but just to the one side of it, so the light from the star falls on a more sensitive part of your eye).

There are a variety of ways to estimate the limiting magnitude, but, in order not to get bogged down into the quagmire of examples, the most effective way is to simply look for the faintest star you can find in the constellation of Ursa Minor. The stars within and surrounding the constellation give you an excellent guide to calculating your NELM. Select the faintest star observable using averted vision and, by referencing that star in a catalog to find its magnitude, the NELM can therefore be established. Once you have used Ursa Minor, it will become a unique reference point all year round for estimating the NELM. The best an observer can hope for is probably +7.0, which will mean the skies are virtually perfect for observing. Take intermittent readings of the NELM and note these down during the course of observations. This will help those collating the data to understand any changes in the clarity of the night sky that the observer has encountered.

With all the necessary details now logged, the observer can await the first meteor. On its appearance, you will need to record the following.

Time of Sighting

Attempt to be as accurate as possible. Other observers from different locations may well be reporting, too, so overall continuity is vital.

Magnitude

You will need to estimate the brightness of the meteor to the nearest whole magnitude. Use nearby stars as a reference point, but only if they occupy a similar elevation above the horizon. Stars lower down or higher up are likely to be affected or by the same token not affected in some way, and this will either distort or enhance their brightness, giving a false impression. As a very crude rule of thumb, the brighter the meteor, the higher the '−' sign. The fainter the meteor, the higher the '+.'

Type

Based on where the meteor appeared, log the sighting as either 'shower' or 'sporadic.' Record the name of the shower, for example, Perseids or Geminids, or simply record 'SP' for sporadic. If unsure, attempt to trace back the line of the meteor to where the radiant is known to lie. If it seems completely at odds with it, it may well be a sporadic. Judging shower meteors against sporadic activity is another part of the learning curve, with frequent observations soon allowing the observers to make clear, concise, and quick identification.

Reliability

Using a scale of A to C, record how well observed the meteor was. Class A represents a meteor that was well observed, right in the line of view, and captured from start to finish; C represents the complete opposite, with the meteor sighted but perhaps from the corner of the eye, possibly incredibly brief, where detail was sketchy. With Class C, if unsure, the observer should not be frustrated about whether it was or was not a meteor, but simply make the decision and move on to the next meteor. During a meteor display, time will be of the essence, and it shouldn't be long before the next meteor needs to be analyzed. Class B represents something in between A and C but, again, make the decision and move on. Tiredness and fatigue can make judging somewhat tricky as the night progresses, so the observer will need to be rather firm when making the call.

Train

During the course of the watch, the observer will encounter a wide spectrum of meteors, some bright, some dim, some fast, some slow, and some that really stand out from the rest. However, all meteors need to have the 'train' described. A lot of the time they will leave no trace, but a proportion of meteors leave a fleeting, glowing wake or train in the night sky, made up of ionized gas. The observer will need to estimate exactly how long the train was visible for before it dissipated and vanished. Naturally, it will be seconds, but try and factor that down into partial seconds if possible. If the observer is fortunate enough, the train may last a while, possibly minutes, but such an occurrence is generally a rarity. Remember, the train is only the residual wake left behind by the meteor, and not the meteor's entire appearance in the sky from start to finish.

Notes

In this section, attempt to describe how fast the meteor was traveling when sighted: 1 = Very Slow, 5 = Fast. Certain showers produce certain types of meteors, and the observer may well find that the bulk of what has been witnessed with a particular shower falls into either the very slow or very fast category. Note down any colors that are apparent, and whether or not there seemed to be a fluctuation in brightness during its appearance. The colors may denote a mix of the chemicals in the meteor's composition.

Other nighttime phenomena are normally fairly recognizable, such as an aircraft and its flashing identification lights, which can attract the observer's gaze for a time. However, as the night progresses and the focus on the task becomes a little hazier, sometimes the observer may have to look twice just to confirm that it is an aircraft, especially if it's near to where you have just seen a meteor. Vapor trails will continue to crisscross the sky, and, on occasion, they can look like the trails left by bright meteors. It is strange, but not unusual, that if an observer is alone the aircraft almost become companions throughout the night, knowing that in our atmosphere, human life exists while all around the observer relative silence in the darkness has descended. By the same token, others consider them a nuisance!

Sky lanterns have become a popular yet, to the observer, annoying piece of entertainment hardware. In the guise of a small hot air balloon, and measuring around a meter in height, these event celebration objects are readily lit and left to drift off into the night air. Although now recognizable, their initial entrance caused much disturbance, first being mistaken for fireballs but more readily as UFOs. The drift and number of objects released tends to be an instant giveaway, but their appearance remains another nighttime phenomena to watch out for.

Occasionally, during your watch, you may witness a fireball. A fireball is an exceptionally bright meteor that attains a magnitude of −3 or above. As a lone observer or part of a group, fireball recording is especially important, but, given its brightness, it is more likely that such a bright meteor will be witnessed by a far greater number of people than those actively dedicated to watching the skies at the time. In recent years, dashboard-mounted cameras in vehicles have sighted many fireballs, with CCTV also picking up not necessarily the fireball itself but the illuminating of the surroundings that the particular camera is covering during the event, almost, in some cases, turning the night to broad daylight in seconds.

Daytime observations of fireballs, though not common, have been recorded, but it is at night, especially during the evening, when most casual witnesses are likely to be involved in seeing the phenomenon. Despite all the technology and possible numbers of witnesses, most fireballs go unseen or, perhaps more frustratingly, are sighted by a solitary individual, perhaps unable to relate what had actually been seen to others. Should a lone astronomical observer be that individual witness, and not someone just casually looking up, remember to record as much detail as possible, using the meteor observing guide as a template. Also, in an attempt to keep the observation as fresh in the mind as possible, don't wait to record it later; do so as soon as possible so the sighting remains fresh. Sketches to illustrate the observation are also desirable.

The meteor observing guide will account for virtually everything that needs to be recorded, but if caught off-guard it is important to at least remember the following:

Date and Time of Observation: Universal Time.

Location: Your exact position, latitude and longitude if possible, or estimated distances from the nearest towns or significant landmarks. Virtually anything to get a fix on your position.

Appearance: Where the fireball appeared in the sky. Which compass point the fireball started at, and the compass point where it faded from sight. Estimate if you lost sight of the fireball at any point by trying to visualize where the path would have taken the fireball.

Coordinates: At night, the elevation in the sky can be determined by stars either known or subsequently referenced in an atlas. Where possible, if familiar with the method, use right ascension (RA) and declination (Dec) to determine its exact position.

Apparent Speed: Using the meteor scale (1 = very slow, 5 = very fast), estimate how quickly the fireball was traveling across the sky. If experienced in observing, this judgment will be more accurate but, all the same, this estimate is another vital piece of information that needs to be gathered. Generally speaking, the appearance of a fireball should not last for more than 10s or so at the very most.

Trains: It needs to be stablished if there was a train left by the fireball. The duration of the train needs to be recorded just as a meteor train would be recorded. Take note of its appearance, color, and, in some cases, the possible inclusion of dust or even smoke.

Sounds: Interestingly enough, sightings of fireballs have been accompanied by sounds, anything from a hissing, whooshing, even crackling, with obvious variants on those descriptions. The mechanism for such noises remains a great source of debate, but if it is clear that the sound has emanated from the fireball then it must be noted, even if the sound is heard when the fireball is no longer visible. It is quite possible that the fireball's passage through the atmosphere creates a sonic boom, heard after its appearance, as shock waves advance towards the observer.

Fragmentation: During its flight, did the fireball seem to break up in any way or did its appearance remain consistent throughout its journey across the sky? It is quite possible that the object gets literally torn apart after entry into Earth's atmosphere. Therefore, although the initial blaze witnessed was that of a fireball, its subsequent fracture into smaller fragments may result in a small shower of meteors. Angle of descent and composition may allow a fireball to remain intact during its entire apparition.

Color: Even a trace or hint of color should be noted, but be clear about what was witnessed. Stick to widely understood colors: red, orange, yellow, green, blue, violet, or white. If the fireball seemed to generate a veritable mix of color, then say something like blue-white or red-orange. Stay away from imaginative colorful labeling.

Magnitude: This will probably give the casual observer the most problems, trying to determine exactly how bright the fireball was. Remember that a magnitude in excess of -3 would class the object as a fireball. The brightness of Jupiter at opposition, -2.9, is a good comparison. Venus at its most brilliant can attain -4.8. A waxing or waning crescent Moon measures -8, a gibbous phase Moon measures -11, with a full Moon measuring -13. If in doubt, use -7 to -9 as a benchmark.

Determining brightness is a skill that sometimes even the most experienced observers can take a little time to deliberate over, but there's no doubt about it that the amount of time watching meteors does train the eye to readily distinguish in seconds the level of brightness.

When a meteor reaches a certain brightness there is some debate over exactly how to classify it, with the observer then becoming dependent on other reports, made by both human and, in some cases, electronic eyes. Identifying a fireball is tricky in itself, but, because of its apparent brightness, the seasoned observer instantly knows that it is not a 'standard' meteor. However, should the brightness exceed magnitude −9, then the fireball's classification moves up another level, becoming a bolide.

From the Greek *bolis* (meaning "missile" or, more imaginatively, "thrown spear"), a bolide is superior in brightness to a fireball. A bolide may well show some fireball-like traits of brightness, but will overall be noticeably brighter for much longer during its appearance. Therefore, a bolide could be classed as a bright fireball, spectacular in nature, perhaps accompanied by amplified noise emissions or indeed some form of noise that could only come from the bolide (surprisingly that, too, is a great source of debate). Hopefully, by its brightness alone, other observers, both causal and dedicated, would mean that the bolide has been widely witnessed so, although the individual makes a judgment as to what has been seen, it may all come down to a general consensus of opinion.

From a bolide, you can go up a further level to a superbolide, one that is bright enough to be picked up from not just ground level but from space by orbiting satellite sensors. Whereas a bolide would look to exceed a magnitude of −9, a superbolide would exceed −17 and, because of this exceptional brightness, is a very rare sight. The brightness is such that not only may it be accompanied by sounds but may perhaps even cast defined shadows by night and be observable in broad daylight. In such an event, a solitary witness is therefore as unlikely as the superbolide itself. Therefore, although location, time, and general attributes to documenting the event would still be required, there should be more than just a handful of witnesses.

Meteor observing takes in a much broader aspect of astronomy than those areas that target the Moon, planets, and galaxies through the field of an eyepiece. For a start, the observer takes in a great deal more of the sky, with attention on an entire horizon, affording long periods of time scanning a certain aspect of the heavens, which, as the hours pass, continues to visually stimulate as it changes.

The art to observing meteors lies in becoming one with the surroundings and taking in the true majesty of the night sky, a privilege to accompany a passion.

Brian G. Marsden worked further with the data over the coming years, identifying other periods in history that would tie in with the movements of Comet Swift-Tuttle. The comets of A.D. 188 and 69 B.C. corresponded with other orbit data, for example, that of 1862. However, the 1862 return had cast a considerable amount of doubt into astronomers' minds about the comet's orbital calculations as a whole, as there was no accounting for the discrepancy in years when the comet was predicted to return. Therefore, Marsden went back to the visits by a comet in 1737 and 1750, selecting the 1750 visit as being that of Comet Swift-Tuttle and, on that basis, predicting that it would return in 1981.

10. Observing Comets and Asteroids

History has taught us that comets have always attracted attention, often feared but never ignored. On many occasions, a visitation from the depths of space has provoked an unprecedented level of interest, drawing in at times a global audience, with those who rarely cast an eye skywards also caught up in the moment. For the discoverer of a new comet there is instant recognition, with the person's name being attached and an automatic place in a hall of fame that money just can't buy.

The role of amateur astronomers in this particular field of science is significant, with professionals not initially overly keen about such interaction with amateurs. However, as the years have passed, that stance has reversed, with professionals openly encouraging the submissions of observations and reports with the premise that there can never be enough data collected.

The reasons for this change in attitude are many, including the undeniable truth that the amateur/professional gap, though still evident, has narrowed. With the more congenial binding of resources and information between the two having become established, the combination bears fruit as the relationship develops. Another reason for the nurturing of relations is the hardware now at the disposals of amateurs and, probably most significantly, the fact that by pooling all resources available, the number of eyes scanning the heavens can only be more positive and productive, with much more of what is taking place in our skies being documented and recorded on what is now a hitherto unparalleled level.

There are many examples in the long study of cosmic debris of those who have contributed much at amateur level, and in the field of comet hunting the name Jean-Louis Pons has become somewhat legendary. Considering Pons' career began as a door-keeper to the very observatory in which he was promoted

J. Powell, *Cosmic Debris*, Astronomers' Universe,
DOI 10.1007/978-3-319-51016-3_10

following the discovery of a number of comets seems almost fairytale in nature. However, if the discovery of comets wasn't enough, and as if to prove a point, once ensconced at the observatory, Pons flourished as an astronomer, going on to discover 27 comets in total during his working life there. Self-taught, Pons remains a beacon of light and hope for the many who have not embarked on the academic route in astronomy but still feel drawn to the heavens. Pons proved that, given a chance, it is possible that non-academics are capable of making a telling and productive contribution to astronomy. It was people like Pons who championed the amateur, with his peers not only forced to acknowledge his abilities but ultimately also to grant him the status deserved of one so dedicated to his field.

Pons was not a one-off either. Australian William A. Bradfield (1927–2014) discovered his first comet in 1972, going on to discover a total of 18 comets in the two decades that followed. In fact, Bradfield's last comet discovery was made in 2004 at the age of 77. Bradfield also established the record of having all of his finds namely solely after him, with not a single one shared with a co-discoverer—a feat that has never been rivaled in the history of comet hunting. His dedication to finding comets is also worth a tribute. Bradfield's discovery of 18 comets spanned several decades, with the time spent at the eyepiece scanning the heavens totaling 3500 h.

Even with a nine-year period in between finding his penultimate then last comet, Bradfield diligently stuck at the task when perhaps others would have acknowledged that enough was enough. However, the rainy season returned after a drought that lasted from 1995 to 2004 with another find, proving beyond all doubt that patience and perseverance in the field can pay off. It is also worthy of note that Bradfield made one of his discoveries using just a pair of 7 × 35 mm binoculars, once again reinforcing the mission statement that one does not have to own a large, powerful, impressive telescope to make a find.

Lithuanian astronomer and physicist Kazimieras Cernis is not an amateur in his field, but for sheer dedication and determination he must be acknowledged. Having missed out nine times on claiming a comet for himself over a period of eight years observing, one would have thought that this was perhaps something that was just not going to happen, something that was not meant to be. However, with in excess of 800 h spent at the eyepiece, Cernis finally made a discovery of his own and, not only that, went on to discover a further two comets, as if to prove a point. In the case of Cernis and many kindred observers, there does seem to be a passion beyond making just the one discovery, as if driven, a trait reflected by many in the field.

Another significant contribution from the world of amateur astronomy heralds from England, in the form of another prolific comet hunter, George Eric Deacon Alcock (1912–2000). Alcock, born in Peterborough, caught the astronomy bug at the age of eight while witnessing a partial solar eclipse, on April 8, 1921. Alcock's enthusiasm was further inspired by his observations of meteors and the annual meteor showers, a good starting point and eventual platform to build upon for many amateur astronomers. Meteors continue to be as much a catalyst for enthusiasm in the science as they have done for generations, for here one can actually witness interaction first hand between Earth and space, as a piece of cosmic debris darts across the night sky.

By adopting the rather strange but ultimately rewarding technique of memorizing the patterns and formations of thousands of stars, Alcock was able to spot any object in a star field that apparently shouldn't be there. This in itself is quite amazing, with Alcock estimating that he had memorized the positions of 30,000 stars in total!

Following a New Year's resolution on January 1, 1953, Alcock set his sights on discovering a comet, giving himself five years in which to make a find. On August 25,1959, Alcock, engaging his memorization of stars, saw his method bear fruit with the discovery of a comet in the constellation of Corona Borealiz, C/1959 Q1 (Alcock). This was the first comet to be discovered since 1894, when amateur William F. Denning spotted one. Barely five days later, Alcock sighted a second comet, discovered in the morning sky, in the constellation of Cancer. His third find followed in March 1963, his fourth in December 1965, and his fifth in 1983. This fifth discovery was one of his most famous, being tagged in as IRAS-Araki-Alcock, made all the more somewhat unorthodox as it seemed to fit in with just another day in the Alcock household, with George making the discovery after seeing his wife off to bed for the night.

During his time of observing the skies over England, Alcock was to discover five comets and five novae. His first nova (exploding star) discovery was made in April 1967, Nova Delphini, the first British nova to be discovered since 1934. His last nova discovery came at the age of 79, proving that, just like William Bradfield, comet hunters span all ages and, once hooked on hunting, a life dedicated to the pursuit seems a natural course. However, both Alcock and Bradfield were themselves surpassed by Albert F.A.L. Jones (1920–2013), the oldest person to discover a comet at the age of 80 in 2000. His discovery pipped the previous holder for the oldest recorded observer to make a find, Lewis Swift, who discovered his last comet at the age of 79 in 1899.

Alcock, armed with several pairs of binoculars, 15 × 80 and 20 × 105, plus a telescope of moderate size, was another case of showing the world just what an amateur astronomer is capable of given time and dedication. What makes Alcock's case a little more surreal is that his IRAS-Araki-Alcock comet find was made while kneeling on the floor at the top of his stairs, observing through a double-glazed window!

One of the most prolific team efforts for the discovery of comets in the professional world is attributable to David Levy, along with fellow hunters, husband and wife Eugene Shoemaker (1928–1997) and Carolyn Shoemaker (1929–). Working together, the three returned 13 discoveries during their partnership. However, working outside the team, Carolyn has 32 comet discoveries to her credit.

Naming Comets and Asteroids

There can surely be fewer more prestigious accolades than having one's name tagged to a comet or asteroid.

During the nineteenth century, comets had to put in a second appearance before being given a name, with comets that only appeared once simply being designated with a combination of year of discovery, numbers (both Arabic and Roman), and letters. It was not until the twentieth century that comets were routinely named after the discoverer or those who jointly made the discovery, this all being dependent on the time frame attributed to the sequence of events following notification that a new body had been found.

History also reveals some bitterness over disputed claims, with arguments developing over who saw the new find first. Further exceptions exist, the two most notable of these being that of Halley's Comet and Comet Encke, where the discoverer was attributed the find after making calculations about the comet's orbit rather than actually sighting the body in the night sky.

There has been a lot of wrangling over the naming of comets and, in an attempt to impose some hard and fast rules, the IAU set up a separate group to deal with new finds, the Committee on Small Body Nomenclature (CSBN). The CSBN forms a link in the chain that has the following sequence. First and foremost is the sighting itself. This is reported to the IAU, which hands it to the CSBN. Once the CSBN has cleared the new find, it is handed back to the IAU, which forwards it to the Central Bureau for Astronomical Telegrams (CBAT), which then makes the announcement on behalf of the IAU. The following criteria are applied:

- A prefix, alluding to the type of comet, which can be any of the following:
  - P/for a periodic comet;
  - C/for a comet that is not periodic;
  - X/for a comet for which a meaningful orbit cannot be computed;
  - D/for a periodic comet that no longer exists or is deemed to have disappeared.

- The year of discovery.
- An uppercase letter identifying the half-month of observation during that year. The CSBN divides each of the 12 months into two parts, and assigns each half-month a letter. The 24 letters are A–H and J–Y.
- A number representing the order of discovery within that half month.
- For example, C/2001 B1 means it was "1"st "C"omet to be discovered during the second half of January "B" in "2001."

Once the orbital time frame around the Sun has been confirmed by multiple returns, C/2001 B1 will be assigned another name, for example, 1P/Halley or 2P Encke.

There are instances, such as with Comet Shoemaker-Levy 9, where an "X" or "D" instead of "C" has been applied. "X" indicates that the original comet has fragmented and that the fragments are being individually designated a letter. "D" indicates that the comet is no longer in existence, with Shoemaker-Levy 9 starting as C/1993 F2 and ending up as D/1993 F2 after the fragments of the comet plowed into Jupiter.

After the comet's second apparition, the IAU's Minor Planet Center then becomes involved assigning a permanent number indicating the order of the discovery. The designation is then finalized by assigning the comet with the name of its two first discoverers (last name for an individual or one word or acronym for a team of astronomers). The names appear in chronological order and are separated by a hyphen. There are also cases of three names being assigned to a comet, for example, IRAS-Araki-Alcock, but this naming is the exception rather than the norm.

The naming of asteroids takes on a similar format to the naming of comets. Regulated by the IAU, the limiting of names to those of mythological characters, for example Ceres or Pallas, no longer applies. As long as they are not offensive, and have no connection to politics or anything of a military nature, pretty much anything goes. However, a lengthy process can await the assigning of a name to an asteroid, as its orbit needs to be established to a high degree of precision, and this could take some time. When the orbit is determined, the IAU applies a "permanent designation" to the asteroid, a number that is issued in strict numerical sequence. The discoverer is invited to suggest a name for the asteroid, which goes for final approval to a special committee within the IAU. Whereas the naming isn't restricted to mythological characters, the IAU is likelier to approve a name that is from mythology than one that is not. In reality, only a very small proportion have ever been designated a name, perhaps as few as 5% in total.

4179 Toutatis is an example of a named asteroid. It was discovered on January 4, 1989, by Christian Pollas while sifting through photographic plates. Initially, the asteroid had the provisional designation of 1989 AC. The first letter in sequence (A) signifies that it was discovered during the period January 1–15. The second letter (C) shows that it was the third discovery during that period. The letter I is not used, so the remaining 25 letters of the alphabet are used to designate 25 asteroids discovered during each half-month period. When the 25 are all used, the letter code sequence is repeated as many times as is necessary with a numerical subscript now implemented, incremented every 25 discoveries. For example, 1989 AC_1 would indicate the 28th discovery during the first half of January 1989.

Despite being known as 4179 Toutatis, this asteroid is also known as 1934 CT, after its first sighting on February 10, 1934, after which it was lost until its rediscovery in 1989. The '4179' simply denotes an order applied to asteroids, with 4178 Mimeev preceding 4179 Toutatis, and 4180 Anaxagoras following it. However because of the vastly differentiating orbital time frames of some asteroids, the order is not always sequenced by date and year of discovery unless the orbits conform to a sequence that allows this.

The IAU's criteria, which it presents to the discoverer in order to give the asteroid a name, are quite clear and precise. Proposed names must be:

- no more than 16 characters long (including any spaces or punctuation);
- preferably one word;
- pronounceable (in some language);
- written using Latin characters (transliterations of names from languages not written using Latin characters are acceptable);
- non-offensive;
- not identical to or even similar to an existing name of a minor planet or that of natural planetary satellite.

Comets in the Eyepiece

Comets are normally spotted when quite faint and distant, so an optical aid is a necessity. Whereas a pair of binoculars can suffice, as we've discovered in the case of George Alcock, it still remains that, on a generally wider basis of comet hunting, the larger the telescope, the better the chance of finding a new body. Also, the wider the field the telescope can capture the better.

Many amateur astronomers work their way up through binoculars and telescopes, gaining size as they gain knowledge of the particular area of astronomy they become interested in, as varying sizes in telescopes suit different purposes. One who has the ability to afford a very large telescope from the outset should think carefully about the investment, its purpose, and the potential period of ownership. Whereas the telescope is the ultimate symbol linked with astronomy, buying a substantial model when one is only just feeling one's way in astronomy could be considered reckless. Apart from the expense, a waning interest may see a valuable piece of equipment gather dust unnecessarily, so the advice would be to start with a pair of binoculars, which at least if the astronomy bug doesn't catch can be used for other hobbies and interests, whereas a telescope does not afford you the luxury of more readily being used for something else. Be sure that astronomy is the interest for you as, for most, a long but thoroughly engrossing voyage of discovery awaits, with patience, time and dedication the accompanying watchwords (Fig. 10.1).

Make sure that you are certain that comet hunting is the pursuit for you, and that you put in a good number of hours observing comets that have already been charted, in order to get an understanding of their movement across the skies. This time also allows you to know the optical aid being used, and familiarization with whatever is being used as a key element. Most comets are quite faint and can often be mistaken for deep-sky objects, especially elliptical galaxies or nebulae, where the appearance is one of a misty, diffused, fuzzy 'blob,' elongated in the case of elliptical galaxies. It is nothing spectacular to the eye, but merely seeing one will soon overwrite the image of what many perceive a comet to look like, with the thrill of seeing a comet for oneself the overriding pleasure, no matter how non-comet-looking it seems.

Simply getting to know the differences in the appearance of deep-sky objects will determine how potentially fruitful searches can be. Setting a target of finding a comet within a certain time frame can add to the desire to find one but, by the same token, it

Fig. 10.1 More than 100 asteroids were captured in this view from NASA's WISE spacecraft during its primary all-sky survey. Not all of the asteroids are easy to see, but some stand out as a series of dots. The asteroid at center left is called (2415) Ganesa. Image credit: NASA/JPL-Caltech/UCLA (courtesy of NASA)

can also test one's patience to the limit. Try to remember why one is comet hunting, and do not overlook all the other sights in our universe during what is ultimately a journey of discovery that will, merely as a bonus, perhaps allow the finding of a comet. As sound a method as any is to follow the strategy of dedicating oneself to observing just one portion of the night sky so that, without requiring George Alcock's vast memorization of the star patterns, one particular segment is solely concentrated on, with those stars in the field of view engrained in the memory. Within this dedicated area of sky, every star and deep-sky object position would need to be noted, so that anything that appears in the segment of sky that wasn't there previously can be instantly picked up on.

When you do see something, make sure it is not an already charted asteroid or comet. If you are unsure, it may well be prudent to report the find anyway, so that it can be checked. The star field that is being observed will need to be fairly extensive in size, not only to widen the scope for making a comet find but also in order to track movements of any object over a series of nights, with a wide field allowing you to track the more lively of 'movers.'

Comets can start faint and remain relatively faint; some brighten and stay bright for a fair length of time, while others brighten rapidly but may fade as quickly. This unpredictability has kept astronomers guessing as to why such erratic behavior takes place, with such questions and theories only being answered and proven as we have come to understand the composition of comets. Despite this, there remains a significant level of doubt, with a number of comets referred to in this book that simply haven't performed as predicted. It would be wise to be mindful that, to an extent, we are still dealing with a great area of uncertainty with comets. However, that said, to some that makes the art of comet hunting more appealing and more of a challenge.

Apart from the variances in brightness, some comets are condensed, giving the observer a distinct nucleus to concentrate on. Other comets can be quite diffused in nature, rather smudge-like in appearance, where no such clarity exists. Offering a thrill on sighting is any tail seen developing from a comet, one that eventually may end up stretching kilometers across the sky, or the more deceptive one that spans kilometers behind a comet approaching Earth not quite head-on but direct enough so the tail is hidden. A comet may develop two tails or a tail whose appearance changes night after night, or may develop the appearance of a forward 'spike.'

The magnitude of comets is estimated using the same method as estimating the brightness of a variable star, by comparing it against other stars that are constant and of a known brightness. However, this can be a challenge, as the comet's appearance is anything but star-like, especially if quite diffused. However, despite this, there is a way around this complication, and that is by defocusing the comparison stars that are being used as a similarity, making them an approximation to the comet in its size and appearance, whereby a visual match is then established.

Before attempting to evaluate size, brightness, and other elements, the observer needs to be aware of a 'false nucleus.' The false nucleus is a star-like body deep within the coma or central condensation. It is known as 'false' because the true nucleus is hidden from view by sunlight reflecting off dust particles emanating from the true nucleus, our view of which is obscured. There is no rule of thumb here; some comets have a false nucleus, some do not, but observations of a 'false nucleus' are nonetheless valid, so appearance, color, and brightness should all be noted.

Various methods exist for determining brightness, with early work by N. Bobrovnikoff, J.B. Sidgwick, and Max Beyer being used to estimate not just the brightness of the nucleus, but specifically the coma itself. Later, C.S. Morris advanced the work of determining brightness, offering another method for consideration.

There are five established methods:

1. THE VSEKHSVYATSKIJ-STEAVENSON-SIDGWICK (VSS) METHOD (IN-OUT METHOD). This is the most widely used procedure, popularized by J.B. Sidgwick working for the British Astronomical Association during the 1950s.

   The brightness of the comet, now captured and in focus in the eyepiece, is compared with stars that have been defocused until the apparent size is the same diameter as the in-focus coma. The procedure is repeated for several comparison stars to obtain a reliable measurement. It is the comet's mean surface brightness that is estimated. This particular method is best suited to comets that are diffused in nature. Should the comet show any marked difference in brightness (a sharp intensity gradient, strongly condensed, concentrated towards the center), then there may well be difficulty is ascertaining an accurate brightness reading. The observer will need to use a fair amount of judgement, with several tries required until some sort of certainty is established about the comet's brightness.

2. VAN BIESBROECK-BOBROVNIKOFF-MEISEL (VBM) METHOD. This procedure is generally credited to Nicolas T. Bobrovnikoff, but there is evidence to suggest that it may well have been devised much earlier.

   The comet and chosen star are both defocused by the same amount, until each appear more or less similar in size. When both are defocused to what appears to be the same extent, the comet usually still appears to be slightly larger than the star. The brightness of the defocused comet and defocused star are then compared. It is necessary to use several defocused stars to give an accurate comparison. This method is most favorably suited to more condensed comets, where there is a marked difference in brightness towards the center. Similarly, the same result is possible if the observer is bespectacled, as simply removing the spectacles will blur the images of both the comet and the star.

3. BEYER'S METHOD. Devised by amateur astronomer Max Beyer, this method is not entirely dissimilar to that of the VBM method. The difference Beyer placed on ascertaining the brightness is to take the extra-focal procedure to a more extreme level. In order for this method to be effective, the head of the comet must be defocused to many times its in-focus diameter. The optical instrument being used is continually defocused until the comet and stars begin to merge and disappear into the background. The order of the disappearance is then noted. If a chosen star as a comparison disappears before the comet does, it must necessarily be fainter, and indeed vice versa. Measuring disappearance increments between the star and comet, as the focuser is turned, allows the observer to obtain a magnitude value. The selection of several comparison stars is advisable. Beyer's method is best suited to those comets that are highly diffused in nature.

4. MORRIS METHOD (MODIFIED OUT METHOD). The method was formulated by Charles Morris and Stephen James O'Meara during the 1970s, but is commonly referred to as the Morris Method. The observer sufficiently defocuses the comet just enough so that the coma becomes rather uniform in brightness, thus erasing and compensating for the generally steep brightness gradient present in most comets, from outer coma to central condensation. The comparison stars are defocused separately to the same apparent size of the uniform defocused comet image. Several attempts will probably be required in order to master what is generally considered as a difficult method to estimate brightness.

5. THE IN-FOCUS METHOD. This method is long established, has been in use for many centuries, and, upon explanation of how it is applied, one can see why. It involves the simple estimation using the naked eye to compare the brightness of a comet with the surrounding stars, all objects including the comet being in focus. There would appear to be plenty of room for error but, if used in conjunction with other methods, may well assist in gaining a good second opinion.

Measuring the size of the coma is an important aspect of cometary observations. The gauging of the comet's coma diameter not only allows an overall determination of the comet's size but is also a strong and accurate measurement of sky conditions at the time of observation. Poor observing conditions, especially when there is a haze present, or a general loss of a clear and clinical aspect, will make the outer limits of the coma disappear, as it merges into the surrounding fuzziness. This can lead to an underestimation in the size of the coma. There are three established methods:

1. THE USE OF A MICROMETER if at all possible, which should give a precise and accurate determination.
2. MEASURING THE COMA DIAMETER by estimating its length in relation to two stars of known separation. The use of a detailed star atlas would be of great assistance here.
3. THE DRIFT METHOD. This allows the coma to drift across the field of view in the eyepiece. The length of time in seconds, t, for the coma to drift across a north-south aligned cross-hair is measured. The diameter is then calculated from diameter = 0.25 t cos (dec), where 'dec' is the comet's declination. Alternatively, you can time the coma's drift out of the field of view along the east-west line. Stars will drift out of the field of view to the west, with other stars entering the field of view to the east. The diameter is calculated as D = 15 t cos delta where D is the east-west time span.

Probably the simplest way to establish the diameter of the coma is by knowing the field of view of each eyepiece being used. It's not incredibly accurate and does not really count as an established method, but can be used as a starting point, even better if used in conjunction with another method.

Bringing all evaluations together gives the initial premise for the visual aspects of the comet: the nucleus, coma, and tail.

The tail can often be the most interesting aspect of a comet, and establishing the tail and angle is something that perhaps only photography can capture accurately. Comets have two distinctive types of tails: a dust tail and a gas ion or plasma tail. The dust tail contains small, solid particles and forms as sunlight acts on these small particles, gently coaxing them away from the comet's nucleus. Because the pressure from sunlight is relatively weak, the dust particles end up forming a diffused tail with a distinct curvature to its appearance, following a line back from whence the comet came.

A gas ion tail is much more elusive than a dust tail and forms when ultraviolet sunlight tears away one of more electrons from the gas atoms residing in the coma, making them into ions (a process known as ionization). The solar wind then carries these ions straight outwards away from the Sun. The resulting tail, unlike the curved dust tail, is straight and pointed directly away from the Sun, often narrower in appearance. The gas ion tail can also exhibit a beautiful electric blue coloring. Both tails can extend millions of kilometers into space, with the comet moving around the Sun, entering its outward orbit tails-first. Ultimately, as the comet heads out of the Solar System, the Sun's influence on it weakens, with any established tails consequently dissipating and fading.

The tail, whether photographically captured or by the more experienced observer properly sketched, is plotted with the use of a star atlas. The length is measured with a ruler with reference to the scale of the map, or by calibrating the measurement against two stars, the separation of which is accurately known. If there are two tails present and clearly identifiable, both should be recorded separately and not as one entity. The measurement of the angle of the tail is achieved by simply gauging the orientation of the tail relative to the position of the head of the comet, using a protractor, with celestial north being 0°, the east being 90°, and so on. If the dust tail is curved, then take different readings to account for the various angle positions. Making as many observations both photographically and by sketching is vital in the observation of comets, as sky conditions can and do change, with hazier, less clear skies likely to hinder good judgment of a comet's tail.

There are some other comet-related features to look for, but, again, only experience in the field will allow one to make a proper assessment:

ENVELOPES/HOODS. An apparent feature that involves a concentrically placed envelope, encapsulation, or hood around the central condensation.

FANS. Material emanating from the central condensation. This phenomenon is actually as it sounds, with fan-shaped debris.

JETS. These are very radical features, projecting outwards from the central condensation, either as straight lines or as curved phenomena. Jet activity, though still evident in smaller comets, is usually most prominent and active in very large comets. As the general formation is more or less constant, it is prudent to presume that smaller comets still show this activity; it is just not as readily witnessed. Comet Hale-Bopp's jet-like outgassing was not uniformly spread over its nucleus but instead came from several specific jets. By observing these determined jets, astronomers were able to measure, with a good degree of accuracy, the rotational period of the comet, which was found to be around the 11 h 46 min mark. The same principle was applied following extensive observations of Comet Machholz (C/2004 Q2) during February, March, and April 2005.

SPINES. Bright, sharp, narrow streaks seen leading from the central condensation into the dust tail.

RAYS. Intricate strips of light emanating from the central condensation. These strips usually have a white/gray appearance, but blue has also been observed.

SHADOWS OF THE NUCLEUS. A rarity, and not really a shadow, something of an optical illusion. Best captured with sophisticated equipment.

FOUNTAINS. More commonly observed than jets, these features are of a similar nature to the latter, extending at various angles from the central condensation.

KINKS. A feature whereby a bend, twist, or indeed kink is sighted in the comet's dust tail. The solar wind is responsible for this feature, with movement of an individual kink seen traveling downwards through the tail over a period of hours. Kinks can occur quite randomly, with the tail subsequently 'straightening' back out to its original curve following the kink's eventual removal from the length of the tail.

STREAMERS. Fine and intricate thin blue lines, diaphanous and gauze-like in appearance, emanating from the coma and running straight down the gas tail. These streamers can appear as numerous entities, very long in nature, intertwined with the tail. Recorded as a bizarre and intriguing sight, streamers can make for quite a spectacle. NASA's Spitzer space telescope, launched on August 23, 2003, made some interesting observations of streamers. Spitzer, an infrared observatory capable of studying objects both within and far outside our own Solar System, captured a picture of Comet 17P/Holmes. This comet suddenly erupted in November 2007, brightening a million times overnight. Spitzer's picture distinctly shows the comet's outer shell and streamers of dust.

KNOTS. A strange occurrence, whereby a 'lump' or nugget of material is discharged from the coma, making its way downwind through the tail. Also referred to as a sort of dark patch, but distinct enough to be recognized as a random discharge and not an ongoing feature of the dust tail.

Disconnection Events

Where the ion tail, subject to great interference from the solar wind, can become detached from the comet. A sudden switch in solar wind direction can forcibly detach the ion tail from the comet, leaving a distinct and noticeable gap between tail and comet. Such a phenomenon was displayed by Comet Lulin on February 4, 2009. Other comets to have shown a disconnection are Comet Lemmon, 2013, and Comet Encke, 2007. In the case of Comet Encke, a coronal mass ejection (CME) was responsible for the disconnection.

Anti-tails

The anti-tail or 'anti-solar tail' is a perspective trick, which occurs when Earth passes through the plane of the comet's orbit, with the observer on Earth witnessing sunlight reflecting off dust particles that appear to extend in front of the coma pointing towards the Sun. The most famous example of an anti-tail was observed in 1957, with this distinct feature prominent for a time with Comet Arend-Roland.

With cometary observations, patience is the watchword. Simple initial location of a comet is a good starting point, with subsequent monitoring of its progress night after night against the backdrop of stars. This will also allow the observer to become a good judge of cometary brightness against the various stars it passes near to, with the brightness of each individual star adjusting and fine-tuning observations of the comet's own brightness. Learning to distinguish all the various aspects that make up the comet is also key, with photographs and sketches helping to affirm the observations made. Groundwork is essential so that future observations will come naturally, and without too much hesitation, especially if the weather is inclement.

Photographing Comets

The very basic outfit to capture a naked-eye comet consists of a tripod, a remote release (or, if unavailable, a self-timer), and a good digital single-lens reflex camera (DSLR). The camera must have a lens with a focal length of at least 100 mm. The longer the lens, the better the picture should be.

Needless to say, a good, clear view of the sky is essential, away from artificial lighting and light pollution in general. Depending on the comet's position, try also to get a vantage point that positions the comet as far above the horizon as you can. Attaining elevation allows the observer to place the comet under the best possible viewing conditions available.

Set the camera to RAW, which literally produces a raw image, with the image file containing a minimal amount of processed data from the image sensor on the camera. The RAW setting enables the camera to grab as much detail as possible, given such a limited field to work with.

Turn the camera's ISO setting up to at least 800 (the higher the ISO setting, the more sensitive the image sensor). Apart from making the camera more sensitive to what limited light is available, this high ISO setting will increase the shutter speed, which in turn decreases the chances of blurring the image or, if opened long enough, the creation of star trails.

Having placed the camera in manual mode, focus on a bright object in the night sky, then recompose and, ensuring that the camera is in mirror lock up, use the cable release. If you are operating the camera without cable release, follow the same instructions but factor in the self-timer.

Should the comet appear large enough in the sky, good shots can be achieved with a hand-held DSLR and a 300-mm telephoto lens.

Naturally, the best results will be achieved via the use of a telescope, one with a focal length of at least 500 mm or more, coupled with a charge-couple device (CCD). Use the native focal length of the telescope and take a 60-second exposure of the general area of the comet. Having viewed the image or images, select the best frame for the comet and, using polar-aligned tracking, take a series of exposures of about 300–900 s.

In order to process out the image noise associated with all cameras (random variations of brightness or color information in the image), make sure to capture a set of 'dark frame' calibration exposures, using the same settings that were used to take the first set of images but now with the lens covered. Dark frame images will need to be subtracted from your shots before stacking them into one single image.

The advent of the CCD-equipped telescopes and digital technology as a whole in amateur astronomy revolutionized the field, but the sword is double-edged. First, the ability to simply program a telescope to seek out an object without much manual intervention is a time-saver, especially if the weather is inclement. Also, the location, photographing, and sharing of new finds has become virtually instantaneous, with those notified able to quickly turn their attention to the discovery and verify its existence. The need for several days of confirmation and verification no longer exists, and, in a world where the demand for everything is right now, this tailor-made approach will suit many.

However, and rather sadly, it does detract from the simple delight of making the discovery for oneself from scratch. The setting up, alignment, and searching seem to have been lost and made redundant, being described as the way things used to be done. Antiquated and out of date? For some, the assembly of a telescope, use of a star chart, followed by a few nights' patience under clear skies, is immeasurably more rewarding than the 'on a plate' gift of modernization. Somehow, the magic and mystery have been stolen, replaced with the overzealous desire for instant results.

Astronomy is a craft in itself, where the groundwork and learning curve to achieve is spent with the passing of the seasons and changing constellations, the phases of the Moon, and the movements of the planets. Learning to walk before one can run has never been as prevalent as in the astronomical learning curve.

Observing Asteroids

Much of what had been detailed about finding and observing comets applies to asteroids, aside from the fact that an asteroid will appear as a point of light rather than a smudge or a blob. Several hours will be required to ensure that the asteroid is not moving in line with the stars or at the same rate, with a high degree of accuracy in the asteroid's positional required.

Remember to record:

- Date and time to the nearest minute, accounting for local time zones.
- Location of the asteroid using right ascension and declination coordinates.
- Size of telescope and magnification used.
- Any other relevant aspect plus sketches showing the asteroid's position in relation to the nearest stars.

With the use of CCD-equipped telescopes, finding and observing asteroids has never been easier or more accessible, and although like comet hunting there remains a degree of difficulty, the modern age of amateur astronomy has made it much easier. The challenge exists to merely achieve the correct settings on the telescope. That said, like comets, this is true only for known orbital positions; the work still has to be done to find new bodies, something modern technology can only assist with but not deliver.

Given the size of asteroids, a fairly large telescope will be required, with the ability to take several images of a nominated star field during the course of an hour. With images taken then subsequently aligned and alternately displayed in quick succession on a computer screen, any objects moving across the star field can quickly be identified.

Equipped with software that has the stored information of asteroid movements, the task of locating and photographing an asteroid is a relatively easy one but, again, not the finding of new bodies. As with comets, if the observer is searching an area of the sky night after night and an object not cataloged emerges, this could well be a new discovery. However, be careful of other such faint images that might either be a faint galaxy or an anomaly of sorts produced by the CCD imaging.

The doubt over whether or not a new discovery has been made can be dealt with via a computer link up to the Minor Planet Center in Massachusetts, allowing for the relay of information back to the observer as to exactly what asteroids, or indeed comets, are on view at that particular instance in that segment of the night sky. If nothing ties in with what has been sighted, there is a good chance that it is a new find. However, and at the mercy of the weather, a second night's observations will be required to verify the movement of the asteroid, with a second more detailed report then being dispatched to the Minor Planet Center. Confirmation should follow in perhaps as little as a couple of hours as to whether or not it is an undiscovered asteroid, with a designation number then applied. From early formation to designation, completing the cycle of discovery (Fig. 10.2).

Fig. 10.2 Artist's concept of the collisions that have taken place in the asteroid belt resulting in both the creation of lone, solitary asteroids, and groups of asteroids that travel around our Sun as a family (courtesy of NASA)

Searching near the ecliptic will help the observer, as most objects lying in a direction in opposition to the Sun tend to be a little brighter in nature, increasing chances of picking out a new body. Scouring the skies here could well prove more fruitful as opposed to searching elsewhere in the sky.

As with comets, the observer should make his or her initial asteroid-hunting days productive ones, by finding, observing and sketching known asteroids before turning to the actual art of hunting.

11. Endgame

The Sun—A Certain End

If the world waits long enough, it will eventually be hit by a comet or an asteroid. In the future, Earth can expect to be struck by a 500-m-wide asteroid once every 50,000–250,000 years.

Researchers remain at odds to produce a definitive time frame for Earth's eventual obliteration, some stating odds as remote as a strike rate of one hit in the next 300,000 years, others the more alarming 1 in 10 chance of a strike in the next 100 years. Whatever odds come to light, it would be prudent to treat each with respect.

Then there's the possibility of a larger and more substantial impact, which occurs once every 100 million years, by asteroids larger than 10 km across. However, Earth will face an endgame scenario regardless of whether or not in the intervening period our world is impacted by a large rock.

The size of the impact would determine much—the total cessation of life or the creation of a disastrous imbalance, where shifts in the environment and subsequent changes to the climate would yield crop failures and famine, something Earth has had to deal with for centuries, but not on what would be an unprecedented scale.

All good things come to an end, and unless scientists in the future find a way to halt the Sun's own endgame, then this is way civilization will end.

Interfering with the Sun in its twilight years is in itself questionable, as the very fact that it has lived its life is merely a completion of its own cycle, an ending, something that we all face. Meddling with such affairs is perhaps contrary to how the universe actually ticks, almost as if the natural balance is being disturbed. Maybe that boundary line has already been crossed, with a different set of circumstances now in place for the future of Earth that

© Springer International Publishing AG 2017
J. Powell, *Cosmic Debris*, Astronomers' Universe,
DOI 10.1007/978-3-319-51016-3_11

may well have not been there before. The demise of the dinosaurs could well have all been part of the natural order, after which a rebirth took place, the result of which is evident today.

On the other hand, if technology has in the future advanced to the point whereby a measure could be put in place to prolong the Sun's life, then maybe that, too, was meant to be. By the same token, the Sun will continue to be around long enough for the exploration and colonization of other planets, so, as the Sun does reach its endgame stage, Earth might not need saving.

The Sun is a G-type main sequence star, with around half of its 10-billion-year existence having passed. Eventually, it will exhaust its hydrogen and helium. Before this happens, a series of brightening events will take place. First, the Sun will brighten by around 10%, giving the current civilization in residence a climate that now has to deal with a global increase in temperatures. A more damaging increase will follow, with the Sun's luminosity brightening by perhaps a further 40%. Although life on Earth could still exist, anyone who is still around will have to seek refuge within Earth itself, burrowing down as certain desert creatures now do in order to escape the extreme heat. However, life will survive—a survival of the fittest and of those that can adapt. Our sister planet Venus is an inhospitable place, but it is to Venus that the current civilization must look, for that is the future—a runaway climate full of deadly toxins accompanied by searing heat (Fig. 11.1).

As Earth fries, the Sun exits mainstream activity and enters a new phase of its existence as a red giant, expanding as it does so to encompass possibly all of the inner Solar System planets, and it is here that life on Earth will end, but not life elsewhere. Should Earth not be engulfed by the Sun's expansion—and there seems to be scientific evidence to support this notion courtesy of the European Space Agency's Hipparcos satellite—anyone left will see the Sun shrink before yet another expansion phase, this time probably rendering Earth inhabitable, with the Sun's life ending as a white dwarf.

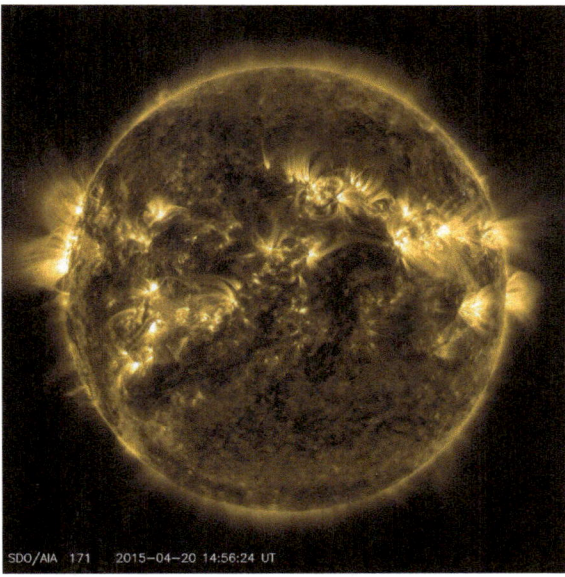

Fig. 11.1 Bright spots and illuminated arcs highlight the more active regions of our Sun. NASA's Solar Dynamics Observatory, April 20, 2015 (courtsey of NASA)

Asteroids and Comets—A Possible End

On the face of it, Earth has been incredibly fortunate inasmuch as the vast majority of cosmic debris so far (the dinosaurs' demise aside) has but scarred Earth in generally remote places, with a minimal loss of life. Scarring from earlier collisions in Earth's past continues, albeit rather rarely, to present itself in modern-day life. One example of this is the discovery of a crater in northwest Scotland. Measuring 40 km across, the crater lies beneath a large area of Scotland, with its center thought to be at Lairg in Sutherland. The impact was discovered after deposits of green molten rock fragments mixed in with red sandstone appeared 'sandwiched' between sandstones that dated back almost 1.2 billion years. This particular crater is only one of 175 known craters, but the most significant found in the UK. With craters over the centuries disappearing as a consequence of both large and small shifts in Earth's tectonic plates, there remain very few noticeable impact sites, aside from the Barringer Crater in Arizona, the Ries

Crater in southeastern Germany, and the Chicxulub Crater in Mexico.

Even with advanced tracking systems to inform us of up-and-coming 'close shaves' for Earth, giving us some edge, it merely serves to confirm the premise that such encounters have been taking place for centuries. The ignorance and, in fairness, inability to comprehend what might strike Earth is a veil that has only in recent times been lifted. As a civilization, there is a potential disaster waiting to befall us from space, but knowledge in this field isn't everything. Granted, a situation presents itself that did not exist before, being a lot more clued-in as to what could hit Earth and what damage it could cause. However, in reality, as the helpless dinosaurs gazed skyward at their impending fate, the same still applies, for, with all the technology and know-how in the world, if a large destructive body has Earth's name on it, it's all over.

However, there is a rather neat counterbalance to this gloomy forecast, insomuch as it probably won't happen. Self-destruction or rampant disease is the more likely predator that stalks us, long before the Sun inhales and exhales its final death-rattle. The threat from NEOs is real, but the odds of a direct hit of a devastating magnitude are greatly stacked in our favor.

Realistically, What Can Be Done to Prevent a Collision?

Quite a fair amount in practice but, while the threat exists, the proactive stance required to tackle such a problem still only exists in the fanciful screen graphics of computer simulations. Although such simulations are welcome, the void that remains between model and actuality is vast. Any such measure to defend Earth would need to be in place long before the realization of a potential collision occurs, commitments that are just not being addressed. Humankind continues to receive many minor encounter wake-up calls but also continues to sideline the issue of defense. However, in fairness, the fact that any global defense has not been implemented is mainly due to the more immediate problems faced by humanity.

In a reasonable span of life such a notion of existence ending in such a manner remains relegated to flights of fancy, not structured in as being a problem worthy of time and thought. Although the thought of life being wiped out from space is entertained, it is merely an historical afterthought of a past event that saw the extinction of the dinosaurs, and look how long ago that was. However, come the time of crisis when an asteroid should appear to be on a direct path to encounter Earth, citizens will readily turn to those 'in the know,' asking questions as to why such a matter has not been addressed, and why money spent on allegedly more trivial and non-life-threatening planetary matters wasn't redirected into saving Earth.

Therefore, some form of justification is necessary, which is why events such as Comet Shoemaker Levy 9 and its impact on Jupiter are crucial. Earth has been hit and is continuing to be struck by rocks, as is our Moon, but in the case of Comet Shoemaker Levy 9, it was a highly visible, modern-era collision witnessed by a global audience with heightened media attention. Jupiter was struck, and if it happened there then the question of it happening on Earth promoted the possibility to a more prominent position.

Therefore, rather than concentrate on a return to the Moon or the ambitious but seemingly not critical effort of putting a human on Mars, attention should really be placed on saving what we already have, Earth, for without that, a return to the Moon or a trip to Mars is irrelevant. Mars has always been considered the next logical target, and there is much to be said in favor of this as, like the Moon, Mars is yet another staging post for deeper space exploration. But one has to ask whether it is totally necessary at this juncture in Earth's evolution, and whether such valuable time and available money would not be better spent on having some form of defense in place, should humanity be called upon to act with some urgency.

Mars will eventually be that step towards spreading humankind throughout space, and avoiding the inevitable demise of the Sun, but not in present times and not for some of reasons that have pushed humans far more quickly in the name of space—to be first and to champion such a mission to another world, ahead of the real importance of going to the red planet.

What Are the Options, Then? Pull It, Push It, Deflect It, or Destroy It?

One of the most promising suggestions is a gravity tractor. The principle is quite straightforward. An unmanned satellite (the gravity tractor) would be launched towards the approaching threat, and, on arrival, the satellite would assume an orbit around the body. The mutual gravitational attraction of the two bodies would allow for the tractor to assert a force on the threat over a period of time, drawing it in whatever direction is required, thus slowly altering its trajectory, enough to safely guide it out of harm's way.

In order for the whole operation to work, careful planning would be required years in advance to orchestrate such a mission, so that the tractor would have enough time to implement this rather casual deflection. Scientists have calculated that a gravity tractor weighing 20 tons would have the capability to deflect a 60-million-ton, 200-m-wide asteroid, given it had at least a year to execute the mission.

Dependent on the size of the asteroid, another consideration would be a more direct use of the satellite's thrusters. Having established an orbit around the body, contact with the threat, then the subsequent application of thrusters, could be applied, gently steering the object into a new course. Indeed, perhaps not just one satellite to perform the maneuver but a 'swarm' of small satellites could be sent to the body. The swarm need not have the capability to attach to the threat beyond landing upon its surface, if each were loaded with an explosive charge to detonate as one at a designated time. Equally, a solution might be the use of the satellite thrusters and no contact, whereby a focused ion beam is produced to prod the body onto a new course.

A more direct route would be the use of kinetic energy or, to put it more bluntly, the delivery of a forceful nudge via the impact of a craft deliberately crashed into the approaching threat. NASA's Deep Impact encounter with Comet Tempel 1 in 2005 demonstrated that such an undertaking could be executed, the probe having released an impactor module into the comet's nucleus in order to examine its composition and structure. The resulting collision between the impactor and the comet was thought to have pushed the Tempel 1 some 10 m off what was then its current trajectory, the 370-kg impactor delivering an equivalent detonation on the nucleus' surface of just under 5 tons of TNT.

A more direct route would be to simply destroy it with nuclear weapons and, given that many NEOs are only a loose collection of rock, the dispersal from such an explosion would sufficiently shatter the approaching threat, rendering it no more than capable of delivering a few sizable meteorite falls and showers of meteors. However, such a radical approach may only make affairs worse, by splitting the body in two, causing two separate threats. Also, an explosion positioned in order to use its blast as a deflection may simply alter the body's course into one that could cause more of a problem. With so much uncertainty to the outcome, is this a reckless approach when so much is at stake? However, along with the gravity tractor and its variants, the use of nuclear weapons would appear to be among the most favored options and, at least in the case of the latter, a constructive rather than negative usage is being applied.

Among the other proposals, the idea of shepherding an asteroid into orbit around Earth to use as our defense is one of the more intriguing suggestions. This asteroid would be dispensed to intersect the trajectory of the threatening body, altering its path into a safer one that does not come into contact with Earth. Equally as intriguing is the use of a large amount of collected light, focused and directed onto the approaching body, creating vaporization and subsequent changes to the body's momentum. The swarm of satellites could also deliver a direct focused beam of light using a collective assembly of mirrors, vaporizing the body by focusing on a specific target area.

Debris or Not Debris

Whatever the future may hold with regard to the vast amount of debris that exists, it is worthy of note that, for much of the time, we happily co-exist with these remnants, the rogues and conformists alike. For this is the overall pattern of existence, the balance, the order from chaos, and the way the universe is and will continue to be.

Fig. 11.2 Embedded in the Carina Nebula is Trumpler 14, an open-cluster containing a family of young, very bright, white-blue stars. Photograph taken by the Hubble Space Telescope (courtesy of NASA)

Humankind has only become aware of the wonders of the night sky for a short while, but, as the advancement of civilization allows the curtain to draw ever wider, our understanding increases —an understanding that will hopefully give a measured response to whatever awaits future generations.

Using the best and most basic intellectual tools at our disposal, and given sufficient time, humankind will learn to adapt as the dawn breaks further on the undiscovered universe from which we were spawned. The advancement and expansion can be within our own governance (Fig. 11.2).

In essence, we are all a great inhalation and the great exhalation, with the universe merely operating on a much larger scale to our lives. Our lives are part of the cosmic debris, as much a part of the fabric as the largest planet or the smallest particle. All encompassing, and as sure as night follows day, the understanding of our belonging is a vital factor in realizing that there is far much more to be learned from our progressive desire to discover and reach out with our minds and see beyond that which can only be physically seen.

Glossary

Ablation A process whereby the atmosphere melts away the surface material of an incoming meteorite

Absolute magnitude Brightness of a star or celestial object if seen from a standard distance of 10 parsecs

Achondrite A stony meteorite lacking chondrules

Albedo The ratio of the light reflected in all directions by a surface to the light incident on it. A perfectly reflecting surface has an albedo of 1, a perfectly absorbing surface has an albedo of 0

Altitude The angular distance between the direction to an object and the horizon. Altitude ranges from 0° for an object on the horizon to 90° for an object directly overhead

Amino acid A carbon-based molecule from which protein molecules are assembled

Amor asteroid A member of a class of asteroids having orbits that cross the orbital distance of Earth

Angular momentum The momentum of a body associated with its rotation or revolution. For a body in a circular orbit, angular momentum is the product of orbital distance, orbital speed, and mass. When two bodies collide or interact, angular momentum is conserved

Annihilation The mutual destruction of a matter-antimatter pair of particles. The charges on the two particles cancel, and the mass of the particles is entirely converted to energy

Annular eclipse A solar eclipse in which the Moon is too far from Earth to block the entire Sun from view and a thin ring of sunlight appears around the Moon

Antimatter A type of matter that annihilates ordinary matter on contact. For every particle, there is a corresponding antimatter particle. For example, the antimatter counterpart of the proton is the antiproton

Aperture The diameter of the main light-gathering lens or mirror, given in inches, centimeters, or meters

© Springer International Publishing AG 2017
J. Powell, *Cosmic Debris*, Astronomers' Universe,
DOI 10.1007/978-3-319-51016-3

Apex   The direction in the sky toward which the Sun is moving. Because of the Sun's motion, nearby stars appear to diverge from the apex

Aphelion   The point in the orbit of a Solar System body where it is farthest from the Sun

Apogee   The point, in an orbit about Earth, that is furthest from Earth

Apollo asteroid   A member of a class of asteroids having orbits that cross the orbital distance of Earth

Apparent magnitude   Brightness of a star or celestial object when observed at its great distance from Earth

Asteroid   A small, planet-like Solar System body. Most asteroids are rocky in makeup and have orbits of low eccentricity and inclination

Asteroid Belt   The region of the Solar System lying between 2.1 and 3.3 astronomical units (AU) from the Sun. The great majority of asteroids are found in the Asteroid Belt

Astronomical unit (AU)   The mean Earth-Sun distance, about 150,000,000 km

Aten asteroid   An asteroid having an orbit with semi-major axis smaller than 1 AU

Atom   A particle consisting of a nucleus and one or more surrounding electrons

Atomic number   The number of protons in the nucleus of an atom. Unless the atom is ionized, the atomic number is also the number of electrons orbiting the nucleus of the atom

Aurora Australis   Light emitted by atoms and ions in the upper atmosphere near the south magnetic pole. The emission occurs when atoms and ions are struck by energetic particles from the Sun

Aurora Borealis   Light emitted by atoms and ions in the upper atmosphere near the north magnetic pole. The emission occurs when atoms and ions are struck by energetic particles from the Sun

Axis   The imaginary line that an object, usually a planet, rotates around

Azimuth   The angular distance between the north point on the horizon eastward around the horizon to the point on the horizon nearest to the direction to a celestial body

Barred spiral galaxy   A spiral galaxy in which the nucleus is crossed by a bar. The spiral arms start at the ends of the bar

Barycenter   The center of mass of a system of bodies

Basalt   An igneous rock often produced in volcanic eruptions

Big Bang   The theory that suggests that the universe was formed from a single point in space during a cataclysmic explosion

Big Crunch   The theory that states that the universe will expand to its maximum point, then contract until it explodes

Black hole   A region of space from which no matter or radiation can escape. A black hole is a result of the extreme curvature of space by a massive compact body

Bolide   A term used to describe an exceptionally bright meteor, possibly accompanied by a sonic boom

Bow shock   The region where the solar wind is slowed as it impinges on Earth's magnetosphere

Brightness   Intensity of light received by an observer from a celestial object

C-type asteroid   One of a class of very dark asteroids whose reflectance spectra show no absorption features due to the presence of minerals

Capture theory   The theory of the origin of the Moon that holds that the Moon formed elsewhere in the Solar System and then was captured into orbit around Earth

Carbonaceous chondrite   A stony meteorite that contains carbon-rich material. Carbonaceous chondrites are thought to be primitive samples of material from the early Solar System

Cassini's division   A conspicuous 1800-km-wide gap between the outermost rings of Saturn

Celestial equator   The circle where Earth's equator, if extended outward into space, would intersect the celestial sphere

Celestial horizon   The circle on the celestial sphere which is 90° from the zenith. The celestial horizon is approximately the boundary between Earth and the sky

Celestial mechanics   The part of physics and astronomy that deals with the motions of celestial bodies under the influence of their mutual gravitational attraction

Celestial pole   The celestial poles are imaginary lines that trace Earth's rotation axis in space

Celestial sphere   An imaginary sphere surrounding Earth. The celestial bodies appear to carry out their motions on the celestial sphere

Centaurs   Small astronomical bodies that generally orbit the Sun between Jupiter and Neptune. Centaurs cross the orbital paths of the major planets

Charge coupled device (CCD)   An array of photosensitive electronic elements that can be used to record an image falling upon it. CCD cameras are composed of silicon chips that are light sensitive, changing detected photons of light into electronic signals that in turn can be used to create images of astronomical objects

Chondrite   A meteorite containing chondrules

Chondrule   A small spherical body embedded in a meteorite. Chondrules are composed of iron, aluminum, and magnesium silicate rock

Chromosphere   The part of the Sun's atmosphere between the photosphere and the corona

Circumpolar stars   Circumpolar stars never set or go below the horizon for observers from specific latitudes

Close pair   A binary system in which the two stars are close enough together that they transfer matter to one another during some stages of their evolution

Cluster of galaxies   A group of galaxies held together by their mutual gravitational attraction

Cluster of stars   A group of stars held together by their mutual gravitational attraction

Coma   A spherical gaseous region that surrounds the nucleus of a comet. The coma of a comet may be 100,000 km or more in diameter

Comet   A small, icy body in orbit around the Sun. When a comet is near the Sun, it displays a coma and a tail

Concretions   A common geologic phenomenon where hard bodies form in sediments before they become sedimentary rocks

Conjunction   The appearance of two celestial bodies, often a planet and the Sun, in approximately the same direction

Constellation   One of 88 regions into which the celestial sphere is divided

Continuous spectrum   A spectrum containing neither emission nor absorption lines

Convection   The process of energy transport in which heat is carried by hot, rising and cool, falling currents or bubbles of liquid or gas

Core   The innermost region of the interior of Earth or another planet

Coriolis effect   The acceleration which a body experiences when it moves across the surface of a rotating body. The acceleration results in a westward deflection of projectiles and currents of air or water when they move toward Earth's equator and an eastward deflection when they move away from the equator

Corona   The outermost layer of the Sun's atmosphere. Gases in the corona are tenuous and hot

Coronal hole   A low density, dim region in the Sun's corona. Coronal holes occur in regions of open magnetic field lines where gases can flow freely away from the Sun to form the solar wind

Coronal mass ejection   A blast of gas moving outward through the Sun's corona and into interplanetary space following the eruption of a prominence

Cosmic background radiation (CBR)   Radiation observed to have almost perfectly uniform brightness in all directions in the sky. The CBR is highly red-shifted radiation produced about a million years after the universe began to expand

Cosmic ray   Extremely energetic ions and electrons that travel through space at almost the speed of light. Most cosmic rays come from great distances and may be produced in supernovas and pulsars

Cosmic string   A tube-like configuration of energy that is believed to have existed in the early universe

Cosmology   The study of the universe as a whole

Crater   A roughly circular feature on the surface of a Solar System body caused by the impact of an asteroid or a comet

Crater density   The number of craters of a given size per unit area of the surface of a Solar System body

Crescent phase   The phase of the Moon at which only a small, crescent-shaped portion of the near side of the Moon is illuminated by sunlight. Crescent phase occurs just before and after a new Moon

Critical density   The value that the average density of the universe must equal or exceed if the universe is closed. If the density of the universe is less than the critical density, the universe will continue to expand forever

Crust   The outermost layer of a planet or satellite

Dark matter   Matter that cannot be detected or has not yet been detected by the radiation it emits. The presence of dark matter can be deduced from its gravitational interaction with other bodies

Declination   The angular distance of a celestial body north or south of the celestial equator. Declination is analogous to latitude in the terrestrial coordinate system

Degree   A unit used to measured angles. There are 360° in a circle

Density   The mass of a body divided by its volume

Differential rotation   Rotation in which the rotation period of a body varies with latitude. Differential rotation occurs for gaseous bodies like the Sun or for planets with thick atmospheres

Differentiation   The gravitational separation of the interior of a planet into layers according to density. When differentiation occurs inside a molten body, the heavier materials sink to the center and the light materials rise to the surface

Doppler effect   The change in the frequency of a wave (such as electromagnetic radiation) caused by the motion of the source and observer toward or away from each other

Dust tail   A comet tail that is luminous because it contains dust that reflects sunlight. The dust in a comet tail is expelled from the nucleus of the comet

Eclipse   The obscuration of the light from the Sun when the observer enters the Moon's shadow or the Moon when it enters Earth's shadow. Also, the obscuration of a star when it passes behind its binary companion

Ecliptic   The plane of Earth's orbit around the Sun. As a result of Earth's motion, the Sun appears to move among the stars, following a path that is also called the ecliptic

Electromagnetic wave   A periodic electrical and magnetic disturbance that propagates through space and transparent materials at the speed of light. Light is an example of an electromagnetic wave

Electron   A low mass, negatively charged particle that can either orbit a nucleus as part of an atom, or exist independently as part of a plasma

Element   A substance that cannot be broken down into a simpler chemical substance. Oxygen, nitrogen, and silicon are examples of the approximately 100 known elements

Ellipse   A closed, elongated curve describing the shape of the orbit that one body follows around another

Elliptical galaxy   A galaxy having an ellipsoidal shape and lacking spiral arms

Elongation   Angular distance of a celestial object from the Sun in the sky

Ephemeris   A tabulation of the positions of a celestial object in sequence for a succession of dates

Equator   The line around the surface of a rotating body that is midway between the rotational poles. The equator divides the body into northern and southern hemispheres

Equatorial jet   The high-speed, eastward, zonal wind in the equatorial region of Jupiter's atmosphere

Equatorial system   A coordinate system, using right ascension and declination as coordinates, used to describe the angular location of bodies in the sky

Equinox   Either of the two points on the celestial sphere where the ecliptic intersects the celestial equator

Escape velocity   The speed that an object must have to achieve a parabolic trajectory and escape from its parent body

Event horizon   The boundary of a black hole. No matter or radiation can escape from within the event horizon

Exosphere   The outer part of the thermosphere. Atoms and ions can escape from the exosphere directly into space

Eyepiece   The lens at the viewing end of a telescope

Fermi paradox   The question that given the known size of the universe, why have we not been contacted and are still alone? Named after Italian physicist Enrico Fermi (1901–1954)

Filament   A dark line on the Sun's surface when a prominence is seen projected against the solar disk

Fireball   An especially bright streak of light in the sky produced when an interplanetary dust particle enters Earth's atmosphere, vaporizing the particle and heating the atmosphere

Focal length   The distance between a mirror or lens and the point at which the lens or mirror brings light to a focus

Focal plane   The surface where the objective lens or mirror of a telescope forms the image of an extended object

Focal point   The spot where parallel beams of light striking a lens or mirror are brought to a focus

Fusion   A nuclear reaction in which two nuclei merge to form a more massive nucleus

Galactic bulge   A somewhat flattened distribution of stars surrounding the nucleus of the Milky Way

Galactic disk   A disk of matter containing most of the stars and interstellar matter in the Milky Way

Galactic equator   The great circle around the sky that corresponds approximately to the center of the glowing band of the Milky Way

Galactic halo   The roughly spherical outermost component of the Milky Way

Galactic nucleus   The central region of the Milky Way

Galaxy   A massive system of stars, gas, and dark matter held together by its own gravity

Gamma ray   The part of the electromagnetic spectrum having the shortest wavelengths

Geosynchronous orbit   An orbit in which a satellite's orbital velocity is matched to the rotational velocity of the planet

Globular cluster   A tightly packed, spherically shaped group of thousands to millions of old stars

Granule   A bright convective cell or current of gas in the Sun's photosphere. Granules appear bright because they are hotter than the descending gas that separates them

Gravitational lens   A massive body that bends light passing near it. A gravitational lens can distort or focus the light of background sources of electromagnetic radiation

Gravity   The force of attraction between two bodies generated by their masses

Great Attractor   A great concentration of mass toward which everything in our part of the Universe apparently is being pulled

Greenhouse effect   The blocking of infrared radiation by a planet's atmospheric gases. Because its atmosphere blocks the outward passage of infrared radiation emitted by the ground and lower atmosphere, the planet cannot cool itself effectively and becomes hotter than it would be without an atmosphere

Habitable zone   The range of distances from a star within which liquid water can exist on the surface of an Earth-like planet

Helioseismology   A technique used to study the internal structure of the Sun by measuring and analyzing oscillations of the Sun's surface layers

Heliosphere   The region of space dominated by the solar wind and the Sun's magnetic field

Hilda asteroids   A group of asteroids with a 3:2 orbital resonance with Jupiter

Hubble's law   The linear relationship between the recession speeds of galaxies and their distances. The slope of Hubble's law is Hubble's constant

Hyperbola   A curved path that does not close on itself. A body moving with a speed greater than escape velocity follows a hyperbola

Igneous rock   A rock formed by solidification of molten material

Inclination   The tilt of the rotation axis or orbital plane of a body

Inertia   The tendency of a body at rest to remain at rest and a body in motion to remain in motion at a constant speed and in a constant direction

Inertial motion   Motion in a straight line at constant speed followed by a body when there are no unbalanced forces acting on it

Inferior planet   A planet whose orbit lies inside Earth's orbit

Infrared   The part of the electromagnetic spectrum having wavelengths longer than visible light but shorter than radio waves

Interferometry   The use of two or more telescopes connected together to operate as a single instrument. Interferometers can achieve high angular resolution if the individual telescopes of which they are made are widely separated

Interstellar matter   Gas and dust in the space between the stars

Ion   An atom from which one or more electrons has been removed

Ionization   The removal of one or more electrons from an atom

Inferior conjunction   A conjunction of an inferior planet that occurs when the planet is lined up directly between Earth and the Sun

Ionosphere   The lower part of the thermosphere of a planet in which many atoms have been ionized by ultraviolet solar photons

Iron meteorite   A meteorite composed primarily of iron and nickel

Isotopes   Nuclei with the same number of protons but different numbers of neutrons

Jets (comets & galaxies)   Venting of gas from weakened areas of a comet's nucleus. Also, a narrow beam of gas ejected from a star or the nucleus of an active galaxy

Kardashev scale   A method of measuring a civilization's level of technological advancement, formulated by Russian astronomer Nikolai Kardashev

Kepler's laws of planetary motion   Three laws, discovered by Kepler, that describe the motions of the planets around the Sun

Kinetic energy   Energy of motion. Kinetic energy is given by one half the product of a body's mass and the square of its speed

Kirkwood gaps   Regions in the Asteroid Belt where a decreased number of asteroids are found, possibly the result of gravitational interactions with Jupiter. Named after astronomer Daniel Kirkwood (1814–1895), who first observed these gaps

Kuiper Belt   A region beyond Neptune within which a large number of comets are believed to orbit the Sun. Short period comets are thought to originated in the Kuiper Belt

Lagrangian points   Positions in an orbital configuration where a small body, under the gravitational influence of two larger ones, will remain approximately at rest relative to them. Named after 18th century

Italian astronomer and mathematician Joseph-Louis Lagrange (1736–1813)

Latitude   The angular distance of a point north or south of the equator of a body as measured by a hypothetical observer at the center of a body

Lava   Molten rock at the surface of a planet or satellite

Libration points   See Lagrangian points

Light   The visible form of electromagnetic radiation

Light curve   A plot of the brightness of a body versus time

Light year   A unit of length equal to the distance that light travels in one year in a vacuum, about 9.46 trillion km

Limb   The apparent edge of the disk of a celestial body

Lithosphere   The rigid outer layer of a planet or satellite, composed of the crust and upper mantle

Local Group   The small cluster of galaxies of which the Milky Way is a member

Longitude   The angular distance around the equator of a body from a zero point to the place on the equator nearest a particular point as measured by a hypothetical observer at the center of a body

Luminosity   The rate of total radiant energy output of a body

Luminosity class   The classification of a star's spectrum according to luminosity for a given spectral type. Luminosity class ranges from 'I' for a supergiant to 'V' for a dwarf (main sequence star)

Luminosity function   The distribution of stars or galaxies according to their luminosities. A luminosity function is often expressed as the number of objects per unit volume of space that are brighter than a given absolute magnitude or luminosity

Lunar eclipse   The darkening of the Moon that occurs when the Moon enters Earth's shadow

M-type asteroid   One of a class of asteroids that have reflectance spectra like those of metallic iron and nickel

Magellanic Clouds   Two irregular galaxies that are among the nearest neighbors of the Milky Way

Magma   Molten rock within a planet or satellite

Magnetosphere   The outermost part of the atmosphere of a planet, within which a very thin plasma is dominated by the planet's magnetic field

Magnitude   A number, based on a logarithmic scale, used to describe the brightness of a star or other luminous body. Apparent magnitude describes the brightness of a star as we see it. Absolute magnitude describes the intrinsic brightness of a star

Mantle   The part of a planet lying between its crust and its core

Maria   A dark, smooth region on the Moon formed by flows of basaltic lava

Mass   A measure of the amount of matter a body contains. Mass is also a measure of the inertia of a body

Maunder minimum   A period of few sunspots and low solar activity that occurred between 1640 and 1700

Mean solar time   Time kept according to the average length of the solar day

Meridian   The circle on the celestial sphere that passes through the zenith and both celestial poles

Mesosphere   The layer of a planet's atmosphere above the stratosphere. The mesosphere is heated by absorbing solar radiation

Messier Objects   List of deep sky objects compiled by Charles Messier (1730–1817)

Metallic hydrogen   A form of hydrogen in which the atoms have been forced into a lattice structure typical of metals. In the Solar System, the pressures and temperatures required for metallic hydrogen to exist only occur in the cores of Jupiter and Saturn

Metamorphic rock   A rock that has been altered by heat and pressure

Meteor   A streak of light produced by a meteoroid moving rapidly through Earth's atmosphere. Friction vaporizes the meteoroid and heats atmospheric gases along the path of the meteoroid

Meteor shower   A temporary increase in the normal rate at which meteors occur. Meteor showers last for a few hours or days and occur on about the same date each year

Meteorite   The portion of a meteoroid that reaches Earth's surface

Meteoroid   A solid interplanetary particle passing through Earth's atmosphere

Microlensing event   The temporary brightening of a distant object that occurs because its light is focused on Earth by the gravitational lensing of a nearer body

Micrometeorite   A meteoritic particle less than a 50 millionths of a meter in diameter. Micrometeorites are slowed by atmospheric gas before they can be vaporized, so they drift slowly to the ground

Milky Way   The galaxy to which the Sun and Earth belong. Seen as a pale, glowing band across the sky

Mineral   A solid chemical compound

Minor Planet   Another name for asteroid

Molecular cloud   A relatively dense, cool interstellar cloud in which molecules are common

Momentum   A quantity, equal to the product of a body's mass and velocity, used to describe the motion of the body. When two bodies collide or otherwise interact, the sum of their momenta is conserved

Near Earth Asteroid (NEA) Near Earth Object (NEO)   Bodies who orbits come into close proximity with Earth

Neutral gas   A gas containing atoms and molecules but essentially no ions or free electrons

Neutrino   A particle with no charge and probably no mass that is produced in nuclear reactions. Neutrinos pass freely through matter and travel at or near the speed of light

Neutron   A nuclear particle with no electric charge

Neutron star   A star composed primarily of neutrons and supported by the degenerate pressure of the neutrons

Neutronization   A process by which, during the collapse of the core of a star, protons and electrons are forced together to make neutrons

North celestial pole   The point above Earth's North Pole where the polar axis, if extended outward into space, would intersect the celestial sphere

Nova   An explosion on the surface of a white dwarf star in which hydrogen is abruptly converted into helium

Nucleus   An irregularly shaped, loosely packed lump of dirty ice several kilometers across that is a permanent part of a comet

Objective   The main lens or mirror of a telescope

Occultation   An event that occurs when one celestial body conceals or obscures another

Oort Cloud   The region beyond the planetary system, extending to 100,000 AU or more, within which a vast number of comets orbit the Sun. When comets from the Oort Cloud enter the inner Solar System, they become new comets

Opposition   The configuration of a planet or other body when it appears opposite the Sun in the sky

Orbit   The elliptical or circular path followed by a body that is bound to another body by the two bodies' mutual gravitational attraction

Organic molecule   A molecule containing carbon

Oscillating universe   A theory that the universe goes through continual phases of expansion and contraction

Outgassing   The release of gas from the interior of a planet or satellite

Ozone   A molecule consisting of three oxygen atoms. Ozone molecules are responsible for the absorption of solar ultraviolet radiation in Earth's atmosphere

Parabola   A geometric curve followed by a body that moves with a speed exactly equal to escape velocity

Parallax   The shift in the direction of a star caused by the change in the position of Earth as it moves around the Sun

Parsec   A unit of distance equal to about 3.26 light-years

Penumbra   The outer part of the shadow of a body where sunlight is partially blocked by the body

Perigee   The point, in an orbit around Earth, that an object is nearest to Earth

Perihelion   The point in the orbit of a body when the body is closest to the Sun

Perturbation   A deviation of the orbit of a Solar System body from a perfect ellipse due to the gravitational attraction of one of the planets

Photon   A massless particle of electromagnetic energy

Photometry   The measurement of the light emitting from astronomical objects

Photosphere   The visible region of the atmosphere of the Sun or another star

Planetesimal   A primordial Solar System body of intermediate size that accreted with other planetesimals to form planets or satellites

Plasma   A fully or partially ionized gas

Plasma tail   A narrow, ionized comet tail pointing directly away from the Sun

Potentially hazardous asteroid (PHA)   Group of asteroids that carry a collision potential with Earth

Precession   The slow, periodic conical motion of the rotation axis of Earth or another rotating body

Prominence   A region of cool gas embedded in the corona. Prominences are bright when seen above the Sun's limb, but appear as dark filaments when seen against the Sun's disk

Proper motion   The rate at which a star appears to move across the celestial sphere with respect to very distant objects

Protein   A large molecule, consisting of a chain of amino acids, that make up the bodies of organisms

Proton   A positively charged nuclear particle

Protostar   A star in the process of formation

Pulsar   A rotating neutron star with beams of radiation emerging from its magnetic poles. When the beams sweep past Earth, we see "pulses" of radiation

Quantum mechanics   The branch of physics dealing with the structure and behavior of atoms and their interaction with light

Quasar   A distant galaxy, seen as it was in the remote past, with a very small, luminous nucleus

Radial Velocity   The part of the velocity of a body that is directed toward or away from an observer. The radial velocity of a body can be determined by the Doppler shift of its spectral lines

Radiant   The point in the sky from which the meteors in a meteor shower seem to originate

Radio galaxy   A galaxy that is a strong source of radio radiation

Radioactivity   The spontaneous disintegration of the unstable nucleus of an atom

Reflectivity   **The ability of a surface to reflect electromagnetic waves. The reflectivity of a surface ranges from 0% for a surface that reflects no light to 100% for a surface that reflects all the light falling on it**

Reflector   **A telescope in which the objective is a mirror**

Refractor   A telescope in which the objective is a lens

Regolith   The surface layer of dust and fragmented rock, caused by meteoritic impacts, on a planet, a satellite, or an asteroid

Resolution   The ability of a telescope to distinguish fine details of an image

Resonance   The repetitive gravitational tug of one body on another when the orbital period of one is a multiple of the orbital period of the other

Retrograde motion   The westward revolution of a Solar System body around the Sun

Right ascension (RA)   The angular distance of a body along the celestial equator from the vernal equinox eastward to the point on the equator nearest the body. Right ascension is analogous to longitude in the terrestrial coordinate system

Roche limit or Roche radius   The distance from a planet or other celestial body within which tidal forces from the body would disintegrate a smaller object. Term formulated by French mathematician Édouard Roche (1820-1833)

S-type asteroid   One of a class of asteroids whose reflectance spectra show an absorption feature due to the mineral olivine

Sedimentary rock   A rock formed by the accumulation of small mineral grains carried by wind, water, or ice to the spot where they were deposited

Search for Extra-Terrestrial Intelligence (SETI)   NASA-led project to search for signs of extra-terrestrial intelligence

Seismic wave   Wave that travels through the interior of a planet or satellite and is produced by an earthquake or its equivalent

Sidereal clock   A clock that marks the local hour angle of the vernal equinox

Silicate   A mineral whose crystalline structure is dominated by silicon and oxygen atoms

Solar constant   The solar energy received by a square meter of surface oriented at right angles to the direction to the Sun at Earth's average distance (1 AU) from the Sun. The value of the solar constant is 1372 watts per square meter

Solar flare   A brief, sudden brightening of a region of the Sun's atmosphere, probably caused by the abrupt release of magnetic energy

Spectral class   A categorization, based on the pattern of spectral lines of stars, that groups stars according to their surface temperatures

Spectrograph   A device used to produce and record a spectrum

Spectroscopy   The recording and analysis of spectra

Spicule   A hot jet of gas moving outward through the Sun's chromosphere

Spiral arm   A long, narrow feature of a spiral galaxy in which interstellar gas, young stars, and other young objects are found

Spiral galaxy A flattened galaxy in which hot stars, interstellar clouds, and other young objects form a spiral pattern

Star A massive gaseous body that has used, is using, or will use nuclear fusion to produce the bulk of the energy it radiates into space

Starburst galaxy A galaxy in which a very large number of stars have recently formed

Steady state theory A cosmological theory in which the universe always remains the same in its essential features, such as average density. In order to maintain constant density while expanding, the steady state theory required the continual creation of new matter

Stellar occultation The obstruction of the light from a star when a Solar System body passes between the star and the observer

Stellar parallax The shift in the direction of a star caused by the change in the position of Earth as it moves around the Sun

Stellar population A group of stars that are similar in spatial distribution, chemical composition, and age

Stony meteorite A meteorite made of silicate rock

Stony-iron meteorite A meteorite made partially of stone and partially of iron and other metals

Stratosphere The region of the atmosphere of a planet immediately above the troposphere

Sublimation The change of a solid directly into a gaseous state

Sunspot A region of the Sun's photosphere that appears darker than its surroundings because it is cooler

Sunspot cycle The regular waxing and waning of the number of spots on the Sun. The amount of time between one sunspot maximum and the next is about 11 years

Sunspot group A cluster of sunspots

Superior conjunction A conjunction that occurs when a planet passes behind the Sun and is on the opposite side of the Sun from Earth

Supernova An explosion in which a star's brightness temporarily increases by as much as 1 billion times. Type I supernovas are caused by the rapid fusion of carbon and oxygen within a white dwarf. Type II supernovas are produced by the collapse of the core of a star

Synchronous rotation Rotation for which the period of rotation is equal to the period of revolution. An example of synchronous rotation is the Moon, for which the period of rotation and the period of revolution around the Earth are both 1 month

Synodic month The length of time (29.53 days) between successive occurrences of the same phase of the Moon

Synodic period The length of time it takes a Solar System body to return to the same configuration (opposition to opposition, for example) with respect to Earth and the Sun

Tektite A small, glassy material formed by the impact of a large body, usually a meteor or an asteroid

Terminal velocity The speed with which a body falls through the atmosphere of a planet when the force of gravity pulling it downward is balanced by the force of air resistance

Thermosphere The layer of the atmosphere of a planet lying above the mesosphere. The lower thermosphere is the ionosphere. The upper thermosphere is the exosphere

Transverse velocity The part of the orbital speed of a body perpendicular to the Sun between the body and the Sun

Trojan asteroid One of a group of asteroids that orbit the Sun at Jupiter's distance and lie 60° ahead of or behind Jupiter in its orbit

Troposphere The lowest layer of the atmosphere of a planet, within which convection produces weather

Ultraviolet The part of the electromagnetic spectrum with wavelengths longer than X-rays but shorter than visible light

Umbra The inner portion of the shadow of a body, within which sunlight is completely blocked

Universe All matter and space

V-type asteroid The asteroid Vesta, which is unique in having a reflectance spectra resembling those of basaltic lava flows

Van Allen Belts Two doughnut-shaped regions in Earth's magnetosphere within which many energetic ions and electrons are trapped

Velocity A physical quantity that gives the speed of a body and the direction in which it is moving

Visual binary star A pair of stars orbiting a common center of mass in which the images of the components can be distinguished using a telescope and which have detectable orbital motion

Wavelength The distance between crests of a wave. For visible light, wavelength determines color

WIMPS Weakly interacting massive particles, 10–100 times the mass of a proton

Wormhole A speculative feature of a black hole that supposedly connects our universe with another universe

X-ray The part of the electromagnetic spectrum with wavelengths longer than gamma rays but shorter than ultraviolet

X-ray burst Sporadic burst of x-rays originating in the rapid consumption of nuclear fuels on the surface of the neutron star in a binary system

Zenith The point on the celestial sphere directly above an observer

Zodiacal constellations The band of constellations along the ecliptic. The Sun appears to move through the twelve zodiacal constellations during a year

Zodiacal light The faint glow extending away from the Sun caused by the scattering of sunlight by interplanetary dust particles lying in and near the ecliptic

Zonal winds   The pattern of winds in the atmosphere of a planet in which the pattern of wind speeds varies with latitude

Zone of convergence   According to plate tectonics, a plate boundary at which the crustal plates of a planet are moving toward one another. Crust is destroyed in zones of convergence

Zone of divergence   According to plate tectonics, a plate boundary at which the crustal plates of a planet are moving away from one another. Crust is created in zones of divergence

Index

A

Advanced Composition Explorer, 17
Aerolites, 96
Aetius, Falvius, 55
Agrippa, Marcus Vipsanius, 54
Alcock, George, 81, 223, 225
Alinda family, 126
al-Jawzi, Ibn, 57, 59
Amor, 32, 35
Andromedids, 201–204
Anti-tail, 45, 73, 79, 232
Apollo 13, 77, 78
Araki, Genichi, 81
Arend, Sylvain, 73
Aristotle, 45–47
Armagh Observatory, 200
Arouet, Francois-Marie, 36
Asteroids
    C-type, 26, 27, 122, 124
    M-type, 27
    S-type, 26, 27, 29, 32–34, 124, 125, 131

B

Babylonian, 46, 51, 53
Barnard, Edward Emerson, 69
Bayeux Tapestry, 58, 59
Beech, Martin, 110
Beila, Wilhelm von, 50, 202
Beljawsky, Sergei Ivanovich, 68
Bennett, John ('Jack') Caister, 77
Bennu, 127, 128
Berolina, 422, 32
Bessel, Friedrich, 49
Big Bang, 1, 7, 9, 16
Biot, Édouard, 57
Bolide, 109, 113, 117, 181, 215
Bopp, Thomas, 85
Bradfield, William A., 218, 220
Brahe, Tycho, 45–47
Brooks, William Robert, 68
Brown, Michael E., 13
Burnham, Robert Jr., 74

C

California Institute of Technology (Caltech), 3
Callisto, 30
Canada-France-Hawaii Telescope, 12
Centaurs, 19, 41, 42
Central Bureau for Astronomical Telegrams (CBAT), 86, 221
Ceres, 19, 22, 23, 25, 31, 222
Chaldean, 46, 47
Chalons, Battle of, 55
Charlois, Auguste H., 32
Chassignites, 105, 106
Chassigny meteorite, 106
Chelyabinsk meteor, 113–115, 117, 126, 131
Chladni, Ernst Florens, 100
Chondrites, 34, 37, 97, 98
    ordinary, 34, 98
    carbonaceous, 98, 122
Clocus, 1220, 121, 123
Collisional family, 13, 40
Coma Berenices, 3, 4
Coma cluster, 3, 4
Comet 21P/Giacobini-Zinner, 192
Comet 46P/Wirtanen, 63
Comet 67/P Churyumov–Gerasimenko, 63
Comet 96/P Machholz, 76
Comet Arend-Roland, 72, 73, 232
Comet Beljawsky, 68
Comet Bennett, 77, 78
Comet Brooks, 68
Comet Daniel, 66
Comet De Kock- Paraskevopoulos, 70
Comet Grigg-Skjellerup, 60
Comet Hale-Bopp, 65, 85–87, 231
Comet Halley *See* Halley's Comet
Comet Hyakutake, 82, 83, 85, 88
Comet IRAS-Araki-Alcock, 81
Comet Kohoutek, 77–79
Comet Lexell, 81, 82